RECRUITMENT AND RETENTION OF BLACK STUDENTS IN HIGHER EDUCATION

RECRUITMENT AND RETENTION OF BLACK STUDENTS IN HIGHER EDUCATION

Johnson N. Niba
Regina Norman

Editors

A NAFEO Research Institute Publication Supported By
Grants From
The Carnegie Corporation,
The Charles Stewart Mott Foundation,
The Pew Memorial Trust Company,
and The Rockefeller Foundation.

UNIVERSITY
PRESS OF
AMERICA

Lanham • New York • London

Library of Congress Cataloging–in–Publication Data

Recruitment and retention of Black students in higher education /
Johnson N. niba, Regina Norman, editors.
p.   cm.
"A NAFEO Research Institute publication supported by grants from
the Carnegie Corporation ... [et al.]."
"Co–published by arrangement with the National Association for
Equal Opportunity in Higher Education"– –T.p. verso.
1. Afro–Americans– –Education (Higher) 2. College stutents– –United
States– –Recruiting. 3. College dropouts– –United States. 4. Student
aid– –United States. 5. Afro–American college students. I. Niba,
Johnson N. II. Norman, Regina. III. NAFEO Research Institute (U.S.)
IV. National Association for Equal Opportunity in Higher Education (U.S.)
LC2781.R43   1989   88–38113 CIP
378'.1982– –dc19
ISBN 0–8191–7292–8 (alk. paper).
ISBN 0–8191–7293–6 (pbk. : alk. paper)

All University Press of America books are produced on acid-free paper.
The paper used in this publication meets the minimum requirements of American
National Standard for Information Sciences—Permanence of Paper for Printed Library
Materials, ANSI Z39.48–1984.   ∞

Table of Contents

Acknowledgments
Chapter I

Excerpts from Hickman, C.R. and Silva, M.A., *Creating Excellence*. New York: New American Library, 1984. Reprinted by permission of New American Library, Penguin, Inc.

Excerpt: "The 7-S Framework" from *In Search of Excellence* by Thomas J. Peters and Robert H. Waterman, Jr. Copyright © 1982 by Thomas J. Peters and Robert H. Waterman, Jr. Reprinted by permission of Harper & Row, Publishers, Inc.

Chapter VI

Excerpts from Allen, W.R. Black colleges versus White colleges: The fork in the road for black students. *Change*, 19, May-June, 28-39, 1987. Reprinted with permission of the Helen Dwight Reed Educational Foundation. Published by Heldref Publications, 4000 Albermarle Street, NW, Washington, D.C. 20016. Copyright © 1987.

Excerpts from American Council on Education, Campus presidents lobby Congress. *Higher Education and National Affairs*, 1986, 35, 7, 1 and 12. Reprinted with permission of the American Council on Education, Washington, D.C.

Excerpts from Evangelauf, J. Increases in college tuition exceed inflation rate for seventh straight year. *The Chronicle of Higher Education*, August 12, 1987, 33, 1 & 32. Reprinted with permission of *The Chronicle of Higher Education*, Washington, D.C.

Excerpts from Mooney, C.J. Legislatures' financial support for colleges limited by economic conditions in states. *The Chronicle of Higher Education*, June 10, 1987, 33, 39, 23-24. Reprinted with permission of *The Chronicle of Higher Education*, Washington, D.C.

Excerpts from *U.S. News and World Report*. Paying for College, October 26, 1987, 83-87. Reprinted with permission of the *U.S. News and World Report*, Washington, D.C.

Foreword

Increasingly, retention is becoming the ultimate yard-stick for measuring institutional effectiveness which translates into measuring the extent to which colleges and universities understand and perform their crucial role of helping both academically talented and underprepared students alike to succeed.

Recent studies have determined that aside from testing, admission standards, and financial aid, the factor most responsible for significantly impeding minority access to and successful progress in the collegiate experience is Dropping Out. Consequently, each institution has continued to grapple with its own anti-dropout initiatives and strategies, tailored and developed to suit or meet the particular needs of each respective institution, and aimed at producing the best results.

As the difficulty and cost of recruiting students continue to rise, so does the awareness of the institutions' bottom-line benefit of ensuring the persistence to graduation of each student. In light of this, and considering that the institutions are in the business of educating students, a task that they cannot perform or undertake unless the recruited students, persist in college, improved retention becomes imperative.

Although the crisis is national in scope, the high black dropout rate during a decade of increased minority high school graduation rates, on the one hand attests to how the problem disproportionately affects blacks. On the other hand, while much of the research in the field has concentrated on the documentation of enrollment counts relative to black and white students, dropping out of college, however, has often been depicted as a phenomenon of low socioeconomic class status and poor elementary and secondary education preparation rather than a function of race. For example, a careful analysis of the statistics will indicate that the drop out rate of poor whites is similar to that of minorities.

These six articles will no doubt become useful and timely additions in the slowly but steadily growing body of literature on the subject of retention (persistence to graduation) at all levels of the educational pipeline. The authors all attempt to explore workable solutions and mechanisms for encouraging retention and combating the problem of dropping out.

Chapter I

Creating and Achieving Excellence: Laying the Foundation for Successful Recruitment and Retention of Black Students

Dr. Harold W. Lundy
Vice President for Administration
and Strategic Planning
Grambling State University
and
Dr. Glenda F. Carter
Coordinator of Research
Grambling State University

Introduction

In recent years, black student recruitment and retention has become an issue of nation-wide concern in the
wake of significant declines in minority participation in
higher education. As a result, many institutions are step-

ping up efforts to recruit and retain larger numbers of minority students. Various programs, rewards, incentives and other devices are being employed to address the immediate crisis-enrollment situation. In contrast, however, the writers here propose a more in depth, strategic approach to addressing the issue of recruitment and retention of black students at historically black colleges and universities (HBCUs). It is proposed that a foundation for excellence be laid which, in turn, serves as an incubator for the development of successful recruitment and retention strategies.

This chapter outlines one approach to creating and achieving institutional excellence in all programs and activities. The crux of this chapter is that once excellence is attained and sustained via encouraging strategic thinking, building an institutional culture, linking strategic planning with outcomes assessment, modeling institutional effectiveness and integrating strategy and culture, the groundwork is laid for success in all academic, administrative, and service programs and activities. This creation and achievement of excellence guarantees greater success in recruiting and retaining black students on black campuses.

Institutional Strategies

Creating and achieving excellence as an overarching theme can serve as a solid philosophical foundation from which institutional strategies can be derived and implemented to help attain growth and self-sufficiency of the nation's colleges and universities. These strategies would be fundamentally designed to (1) improve academic quality; (2) improve institutional management; and (3) enhance and maintain fiscal stability.

This chapter shows how the adoption of a theme of excellence as a foundation for the development and implementation of institutional strategies can create and achieve excellence in all academic programs and activities. It focuses on the measurement of organizational excellence using both objective and subjective yardsticks. Some objective procedures which can be used to evaluate excellence include various outcomes such as:

1. student scores on standardized examinations or locally constructed examinations;
2. performance of graduates in graduate school;
3. pre- and post-testing of students;
4. achievement in general education;
5. performance of graduates of professional programs on licensure examinations;
6. placement of graduates of occupational programs in positions related to their fields of preparation;
7. financial soundness and stability;
8. adequacy and use of institutional resources (assets);
9. rate of job placement for graduates; and
10. rate and quality of placement in graduate or professional education.

At the other end of the scale, some subjective criteria for gauging excellence of an institution's programs include:

1. quality of management;
2. quality of products and services;
3. presence of innovation;
4. "values" as a long-term investment;

5. ability to attract, develop and keep talented people;
6. community and environmental responsibility;
7. perceptions of growth toward objectives;
8. peer evaluation of educational programs;
9. structured interviews with students and graduates;
10. changes in students' values as measured by standard instruments or self-reported behavior patterns;
11. opinion surveys of students, graduates, and dropouts regarding program quality;
12. opinion surveys of employers of graduates regarding job-related competence of graduates;
13. style of governance;
14. regional and national program rankings; and,
15. external recognitions (honors, awards) achieved by students or graduates;

These objective and subjective criteria are not intended to be all-inclusive or mutually-exclusive. They are merely representative of a practical approach to creating and achieving excellence, to outcomes assessment, and to quality assurance.

However, before the objective and subjective criteria can be applied to assess excellence (outcomes), it is essential that they be grounded in a strong foundation—a foundation which espouses the viewpoint excellence can be established by pursuing four important strategies:

1. Encouraging strategic thinking throughout the university's organizational structure;
2. Building an institutional (corporate) culture;
3. Linking strategic planning with outcomes assessment; and

4. Implementing a model of institutional effectiveness (adoption of the 7-S Framework as a planning paradigm).

The contribution of each of these strategies to creating and achieving excellence is presented and discussed in this chapter.

The Foundation of Excellence: Strategy and Culture

The foundation for excellence is rooted in the institution's strategies for success. The leadership of an organization must understand that it is very difficult for that enterprise to succeed without laying a strong foundation of strategic thinking and culture-building. Figuratively speaking, the laying of this foundation can be compared to "a man who built a house, and dug deep, and laid the foundation on a rock; and when a flood arose, the stream beat vehemently upon that house, and could not shake it; for it was founded upon a rock." Likewise, an organization built upon a solid foundation (rock) can successfully withstand the challenges to its survival, continuity, and perpetuity; and certainly, attracting and retaining students is one of the major challenges.

Endemic to this whole notion of building a foundation for excellence is the development of a strategy and the building of an institutional culture. A detailed and thorough discussion of these concepts are not presented here since the chapter is primarily concerned with practical implementation. However, we do present some essential elements which help in understanding the concepts. The following working definitions of strategy and culture are presented in order to provide for unity of thought among the readers.

1. Strategy—the hard-nosed American approach to business that traditionally stresses the impact of competitive advantage on the operating results (bottom line) of the enterprise.
2. Culture—the careful attention to organizational and people needs (an approach for which many Americans greatly admire the Japanese). (Hickman & Silva, p. 25).

Practical implementation of these concepts requires that they be synthesized after a vision of the organization's future has been developed. After the firm's vision of the future has been ascertained, it is necessary that strategies be implemented to make that vision a reality. Next, the organization will need to nurture a corporate culture that is motivated and supported by and dedicated to the vision.

The important linkage of strategic thinking and corporate culture-building demands that leaders not only cultivate a broad vision but master the skills to implement that vision. These leaders must have the ability to see "crisis as opportunity, not danger, and create a future equally responsive to operating results (outcomes assessment) and to an organization's future" (Hickman & Silva, p. 26). Thus, attracting and retaining students at HBCUs necessitates that skills, talents and other resources be developed as well. The vision is not enough.

Encouraging Strategic Thinking

Building a foundation for excellence demands strategic thinking. It is regrettable, however, that the prevailing sentiment seems to be that more planning is better planning. As a result, many organizations have over-gathered and over-analyzed mountains of strategic data to ensure success. Unfortunately, their planning processes have generated static and voluminous documents which are very seldom used to chart the future course of these institutions.

Colleges and universities cannot be obsessed with formal, complex planning. For many institutions, enlightenment has come via a revolution in thought prompted by William Ouchi's (1981) influential book *Theory Z*. From Ouchi's book, we have learned that *strategic thinking*, not complex planning processes, creates excellence. Therefore, it is imperative that institutional line managers be given more responsibility for strategy.

Actions based on strategic thinking must effectively satisfy customer (student) needs, gain a substantial advantage over competitors, and capitalize on institutional strengths. It must be emphasized, even at the risk of being repetitive, that locating, attracting, and retaining customers (students) is the *purpose* and *essence* of strategic thinking.

Unfortunately, locating, attracting and retaining clientele is not as easy as some success stories would suggest. To create excellence through strategic thinking, colleges and universities must establish adaptive subsystems to deal more effectively with three major aspects of their organizational structure: (1) their customers; (2) their competitors; and, (3) their own internal environments. The actions that an institution must take to achieve mastery over these aspects are described in Table 1.

In encouraging strategic thinking, the executive management must continuously stress to all planning/budgeting units that strategic management is a never-ending managerial process of keeping the university constantly ready to seize the very best external opportunities that

TABLE 1

Requirements For Successful Strategies

Strategy Components	Action Colleges Must Take
Clientele	Satisfy client needs, recognizing that different clientele have different needs.
Competitors	Gain a sustainable competitive advantage, keeping product "differentiation" in mind.
Institution	Capitalize on institutional strengths, remembering that it takes time to develop them.

become available to it, while steering the institution away from threats to its continuation. Strategic management is given formal expression and an action orientation through strategic planning. Thinking strategically requires that administrators plan for the institution's future by engaging in the following planning activities:

- analyzing the external and internal environments;
- preparing alternate "scenarios" for dealing with future environments;
- analyzing the competitors' market position;
- understanding the analytical tools of strategic planning;
- capitalizing on the strengths of the institution and mitigating its weaknesses;
- understanding the importance of successful resource deployment; and,
- developing short-term and long-term strategies.

Indeed, an organization's strategic plan, or lack thereof, is an a priori indicator of the depth and quality of top management's thinking about the future and the institution's role in it.

Ironically, the planning processes established in most corporations and institutions often stifle strategic thinking because depending on formalized strategic planning often builds a false sense of confidence. Hence, to revitalize themselves, universities will need approaches which are less complicated, not more so.

By thinking strategically, managers must de-emphasize old skills and develop what have been called "New Age Skills" according to Hickman and Silva (1984) in their book, *Creating Excellence*. The old skills for success were to:

- set goals and establish policies and procedures

- organize, motivate, and control people

- analyze situations and formulate strategic and operating plans

- respond to change through new strategies and organizations

- implement change by issuing new policies and procedures

- get results and produce respectable growth, profitability, and return on investment (p. 30)

The new age skills, however, emphasize:

- *Creative Insight*—asking the right questions;

- *Sensitivity*—doing unto others;

- *Vision*—creating the future;

- *Versatility*—anticipating change;

- *Focus*—implementing change; and,

- *Patience*—living in the long-term (p. 31).

The foundational, integrative, and adaptive skills will greatly assist university managers in achieving excellence in the dynamic future.

Building an Institutional Culture

Every institution must determine the unique basis on which to build its specific institutional culture. For example, at Grambling State University (GSU), the foundational tenet which establishes the basis on which we build our institutional culture is derived from the Statement of Institutional Mission and Philosophy.

A basic tenet of this statement is that "GSU is a place where everybody is somebody." This philosophical statement accentuates the university's commitment to students who have been adversely affected by educational, social and economic deprivation. The university exists to provide opportunities for these students to develop intellectually and to acquire appropriate job skills and self-actualization. Additionally, the university strives to provide equal access to higher education and job opportunities for all applicants, regardless of race, creed, sex, or physical limitations.

It should also be noted that the ". . . everybody is somebody" philosophy extends to the university's treatment of faculty, staff and other constituents. And as previously mentioned, it is the foundational tenet which

10

establishes a sound basis on which to build GSU's *institutional culture.*

By culture-building we mean actions which are aimed at 1) instilling a collective commitment to institutional mission and philosophy, 2) fostering distinctive competence among employees to deliver superior performance, and 3) establishing a consistency that helps attract, retain and develop leaders at all levels. This rather lengthy definition of *culture-building* can be reduced to a simpler expression—selecting, motivating, rewarding, retaining and unifying good employees.

Unfortunately, U.S. business schools have accented hard, quantitative management techniques over supposedly "soft" people skills. Culture-building, however, requires a sharpening of the "soft" people skills, and involves three fundamental steps: (1) *instilling commitment;* (2) *rewarding competence;* and (3) *maintaining consistency.* These three steps to a strong, successful culture and the action that colleges must take to climb these steps are presented in Table 2. A positive, well-defined, established culture inevitably creates other positive rewards. For example, committed employees become enthusiastic recruiters. Committed employees also develop a better rapport with students. The result of these actions is increased enrollment and in all probability, higher retention rates.

GSU's strategy for culture-building is provided as an example. At GSU the Statement of Mission and Philosophy provides the institution's *raison d'être* and its strategic direction and vision. Recently revised to reflect GSU's concern for economic development and entrepreneurial creativity, the Statement of Mission and Philosophy is currently being instilled in new employees and re-instilled in employees of longer tenure.

11

TABLE 2

Three Steps to a Strong Successful Culture

Culture Components	Action Colleges Must Take
Commitment	Install commitment to a common philosophy and purpose, recognizing that employee commitment to the institutional philosophy must coincide with both individual and collective interests.
Competence	Develop and reward competence in key areas, keeping in mind that management will foster greater competence by focusing on one or two key skills at a time rather than by addressing a host of skills all at once.
Consistency	Consistently perpetuate commitment and competence by attracting, developing, and keeping the right people.

Source: Hickman and Silva, 1984, p. 70.

GSU's strategies for enhancing and developing competence include:

1. Formalizing a Management Training and Development Program;
2. Expanding and improving faculty development programs and activities; and,
3. Revitalizing the executive leave and renewal program.

Commitment and consistency will be assured by revising GSU's evaluation system. The revised system will include three components:

1. Adherence to the statements of philosophy;

2. Progress on the achievement of annual objectives (MBO); and,
3. External evaluation of customers/clientele.

Of the three components, we sincerely believe that commitment should be preeminent in GSU's strategic thinking. Therefore, we have developed statements of philosophy for all administrative and support units. These statements of philosophy have been translated into an annual evaluation exercise (Adherence to Statements of Philosophy).

Statements of Philosophy. The purpose of preparing the statements of philosophy was to provide the members of the management team with a common understanding of the *guidelines* for decision-making which are *rooted in GSU's values and principles of behavior.* GSU's statements of philosophy comprise a particular guiding principle from which the university will attain its comparative advantage, and on the side of which, the institution will always err when making decisions regarding policy and tactics. In a very practical sense, these qualitative guidelines are aids to GSU's decision-making, especially in those cases where the choices that the institution has are similar in value and therefore, difficult to make. These guidelines will also prove quite valuable when management must make decisions very quickly, without time for an intense analysis.

GSU's statements of philosophy are also meant to aid in its efforts at performance evaluation. When faced with a question—"Did a person act appropriately in a particular situation or decision?—especially in controversial issues or problems that get out of hand, GSU can refer to the statements of philosophy and determine whether or not the person essentially followed the guidelines which he/she had agreed to follow.

We also believe that GSU's statements of philosophy should be very useful in recruiting. We can often describe to candidates GSU's standards of behavior or principles, but it is far easier to communicate them when they are in writing.

During the initial implementation of our statements of philosophy, the components have been categorized under eight major areas of concern. These categories include:

1. Communications and Customer Relations;
2. Teamwork;
3. Perspective;
4. Professional Development and Training;
5. External Relations;
6. Quality/Cost;
7. Leadership; and,
8. Creating Excellence; Developing Strategy and Building Culture.

Each institution must determine its own strategy for building a strong, successful culture. It must also develop and implement strategies around the culture components which are not important to institutional strategic thinking.

Linking Strategic Planning with Outcomes Assessment

In assuring continuous linkage between strategic planning and outcomes assessment, GSU has begun the implementation of a major Title III program activity— *Achieving Educational Quality Enhancement and Assur-*

ance. The enhancement and assurance will be achieved via the development of predetermined, measurable, competency levels in subject fields and general knowledge and insuring that students attain mastery level skills. Obviously, this activity represents an important strategy for GSU in achieving growth and self-sufficiency by strengthening academic quality. It also represents a linkage of strategic planning with outcomes assessment.

The linkage of strategic planning with outcomes assessment is vital to the effectiveness of any institution. It must be assured. This linkage can be assured by a consistent application of a strategic planning paradigm which depicts three significant determinants of strategy: (1) external opportunities and limitations; (2) internal strengths and weaknesses (internal capabilities); and (3) the institutional culture and values. These three determinants must interact to achieve a "matching process" (match opportunities with capabilities and values). After the "matching process" has been consummated, input is available for the institution to make strategic decisions regarding: (1) mission; (2) goals and objectives; (3) clientele; (4) faculty, programs and services; (5) geographic service areas; and (6) areas of current or potential comparative advantage.

Figures 1 and 1A provide a complete self-explanatory depiction of the linkage between strategic planning and outcomes assessment. This approach to planning relies heavily upon information gained through various assessment programs. A synthesis of this information has led to the development of a Planning and Institutional Research Subsystem which provides valuable decision support to the university's management.

At GSU, outcomes assessment contributes to the evaluation of strengths and weaknesses of existing programs

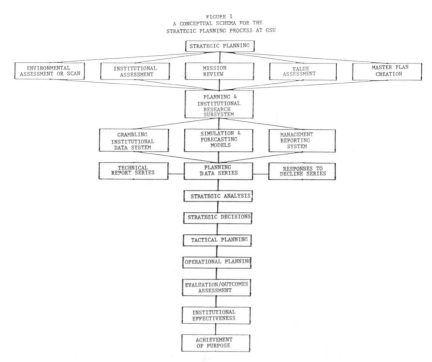

FIGURE 1
A CONCEPTUAL SCHEMA FOR THE
STRATEGIC PLANNING PROCESS AT GSU

and the determination of the need for new programs. Frequently, resources are provided to strengthen programs when assessments reveal qualitative concerns (where resource allocations can improve the quality of the program in question). Additional resources are afforded programs assessed to be capable of achieving national or regional prominence, or those in need of additional resources to sustain such distinctive ratings. Further, assessment results are important in decisions concerning long-term retention or elimination of program, potential for program mergers, and reductions in program size or scope. Outcomes information is among the determinants of candidates for designation as centers of excellence. Assessment information also improves the capability of programs to respond to new external opportunities and to the needs and expectations of those served.

The interrelationships among the activities of strategic planning, departmental planning, academic program reviews, resource allocations, and outcomes assessment reinforce the importance of probability of the success of

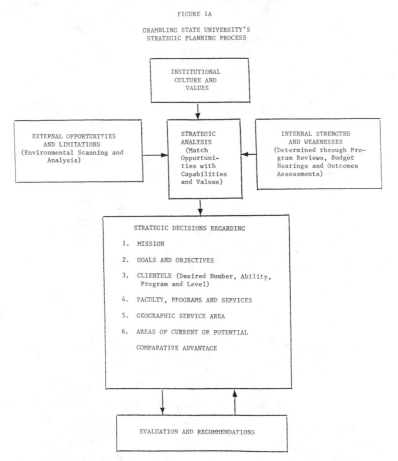

FIGURE 1A

GRAMBLING STATE UNIVERSITY'S
STRATEGIC PLANNING PROCESS

INSTITUTIONAL
CULTURE AND
VALUES

EXTERNAL OPPORTUNITIES
AND LIMITATIONS
(Environmental Scanning and
Analysis)

STRATEGIC
ANALYSIS
(Match
Opportuni-
ties with
Capabilities
and Values)

INTERNAL STRENGTHS
AND WEAKNESSES
(Determined through Pro-
gram Reviews, Budget
Hearings and Outcomes
Assessments)

STRATEGIC DECISIONS REGARDING

1. MISSION

2. GOALS AND OBJECTIVES

3. CLIENTELE (Desired Number, Ability,
   Program and Level)

4. FACULTY, PROGRAMS AND SERVICES

5. GEOGRAPHIC SERVICE AREA

6. AREAS OF CURRENT OR POTENTIAL

   COMPARATIVE ADVANTAGE

EVALUATION AND RECOMMENDATIONS

The strategic planning approach used by GSU employs concepts presented in a workshop sponsored by the National Center for Higher Education Management Systems (NCHEMS) conducted by Dr. Robert Shirley.

17

GRAMBLING STATE UNIVERSITY'S
STRATEGIC PLANNING PROCESS

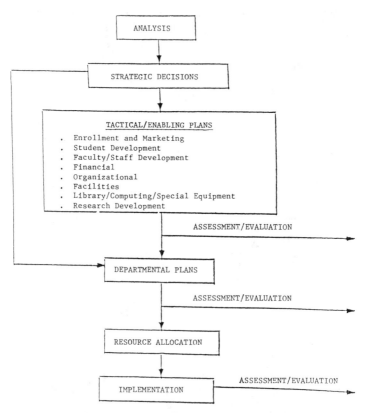

assessment activities, including: (1) the development of goals; (2) the formulation of program and service activities that achieve these goals; (3) the development of a means of evaluating the results; and (4) the sharing of results with the appropriate constituent groups. The information will be valued more because it will be institutionalized in many of the activities which are most important to department heads and faculty. Further, the information gathered can be useful in enhancing departmental reputations, generating additional support for improvement of resources, better relating programs to the needs of those served, and improving understanding of the results of a variety of departmental activities.

Ideally, the most important incentive—for the use of outcomes information should be to achieve the major objectives of individual departments with respect to improvement in teaching, curricula, and ultimately, student performance. Also, outcomes assessment may provide a more positive framework for evaluating and elevating performance levels—a potential which warrants consideration of this approach for measuring both the quality of programs and the effectiveness of resources allocated to bring about programmatic improvements.

Implementing a Model of Institutional Effectiveness: The 7-S Framework as a Planning Paradigm

In 1979, McKinsey and Company introduced a framework for understanding organizational change. Subsequently, the framework was used by others as a useful way of thinking about organizing, as a diagnostic tool, and as a way of pinpointing the cause of organizational malaise.

Presented as Figure 2, the 7-S Framework has the appearance of an atom. The 7-S concept has seven factors, all beginning with the letter "S." These seven factors, defined elaborated upon in the books, *The Art of Japanese Management* and *In Search of Excellence* (Pascale and Athos, 1981; Peters & Waterman, 1982) are:

19

• *Strategy:* plan or course of action leading to the allocation of a firm's scarce resources, over time, to reach identified goals.

• *Structures:* characterization of the organization chart (that is, functional, decentralized, etc.).

• *Systems:* proceduralized reports and routinized processes such as meeting formats.

• *Staff:* "demographic" description of important personnel categories within the firm (that is, engineers, entrepreneurs, M.B.A.s, etc.). "Staff" is not meant in line-staff terms.

• *Style:* characterization of how key managers behave in achieving the organization's goals; also the cultural style of the organization.

• *Skills:* distinctive capabilities of key personnel or the firm as a whole.

• *Superordinate goals* (also called shared values): the significant meanings or guiding concepts that an organization imbues in its members; these goals also represent *the vision, the direction,* or the *mission* of the organization.

The 7-S Framework can be dichotomously utilized in analyzing an organization. An organization may be divided into its hardware—strategy and structure—but also its software—style, systems, staff (people), skills, and shared values.

The McKinsey and Company idea has been used to examine several critical management issues. In *The Art of Japanese Management,* Richard Pascale and Anthony Athos (1981) used the 7-S concept to examine the differences between American and Japanese management. In the book *In Search of Excellence,* Thomas Peters and

FIGURE 2

MCKINSEY'S 7-S FRAMEWORK

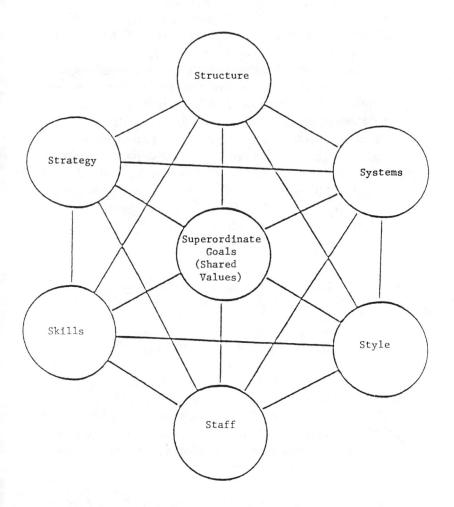

Source: Robert H. Waterman, Jr., Thomas J. Peters, and Julian R. Phillips, McKinsey & Company.

21

Robert Waterman (1982) used the 7-S concept to examine why certain companies succeeded and others fail.

However, colleges and universities can use the 7-S Framework as a planning paradigm and as a management tool for creating and achieving excellence. Planning/budgeting units may be requested to submit their strategic plans in the 7-S Framework. Data provided by these plans can be helpful in allocating scarce resources. Also, the 7-S Framework can be useful in diagnosing the causes of organizational malaise and in formulating programs for improvement.

A basic conclusion about the 7-S Framework is that effective organizational change is really the relationship between structure, systems, style skills, and shared values (superordinate goals). Also, the 7-S Framework provides a sound and pragmatic basis for initiating effectiveness/efficiency improvements.

Strategic Thinking and Culture Building: Working in Tandem

Each of the strategies outlined above are critical to the notion of creating and achieving excellence in all programs and activities. However, the primary foundation of excellence involves a synthesis of strategic thinking and culture-building. A harmonious relationship must exist between these two factors. American and Japanese car manufacturers provide a good illustration of why such a relationship is necessary and how it can be achieved.

Almost all interested observers are now well aware of the fact that the U.S. automobile industry lost its preeminence by believing that "customers will drive whatever we build for them." While at the same time, the Japanese car industry reassessed the automobile's role in a changing global environment and designed cars better suited to that environment. Sadly, we must now admit that Toyota, not GM, dominates the world markets with excellence (Silva & Hickman, p. 26).

In transforming mediocre organizations into excellent ones, there is obviously a need for those firms to become both visionary and realistic, sensitive and demanding, and innovative and practical. This transformation can occur when it is recognized that individuals, not organizations, create excellence.

Individual executives who have developed specific skills create superior organizational performance. Excellence springs from trend-setting levels of personal effectiveness and efficiency. Successful organizations owe their greatness to a few individuals who have mastered leadership skills and passed those skills on to succeeding generations of executives and managers. In grooming future leaders, the mentor makes sure that he passes on both his gift for strategy and his flair for building a strong corporate culture.

To achieve and sustain excellence, business, government and other nonprofit institutions must pay close and simultaneous attention to both strategy and culture, always striving to harmonize them. The strategy-culture matching grid in Figure 3 shows us how to match (unite) strategy and culture. Writers such as Ouchi, Pascale and Athos, Peters and Waterman, and Deal and Kennedy helped us to recognize the importance of culture, but they did not show us how to blend it with strategic thinking. However, Hickman and Silva have shown us how strategic thinking and culture-building work in tandem to create excellence. Let's explore each of the nine combinations shown in Figure 3.

• *Customers/Commitment.* Determine if the organization's collective commitment to a common purpose coincides with the organization's way of satisfying customers' needs.

• *Competitors/Commitment.* Determine if your commitment to a common purpose augments the organization's method for gaining a sustainable advantage over competitors.

• *Company/Commitment.* Determine if your commitment to a common purpose supports the company's attempt to capitalize on its strengths.

• *Customers/Competence.* Determine if the organization's competence to deliver superior performance satisfies the customers' needs.

• *Competitors/Competence.* Determine if your competence to deliver superior performance matches the

FIGURE 3
Matching Strategy and Culture

Culture

Strategy	COMMITMENT	COMPETENCE	CONSISTENCY
CUSTOMERS	Match?	Match?	Match?
COMPETITORS	Match?	Match?	Match?
COMPANY	Match?	Match?	Match

Source: Hickman and Silva, p. 86

24

organization's method of gaining a sustainable advantage over competitors.

• *Company/Competence*. Determine if your competence to deliver superior performance agrees with the organization's efforts to capitalize on its strengths.

• *Customer/Consistency*. Determine if the organization's consistency in perpetuating commitment and competence by attracting and keeping the right people parallel efforts to get and keep customers.

• *Competitors/Consistency*. Determine if consistency is perpetuating the culture agrees with the organization's method of gaining advantage over competitors.

• *Company/Consistency*. Determine if consistently perpetuating the culture enhances the organization's efforts to capitalize on its strengths (Hickman & Silva, 1984, pp. 87–88).

Strategy Plus Culture Equals Excellence

A good way to show how strategy and culture must be matched (alloyed) is to present an equation for excellence:

This equation has three fundamental variables in it:

1. Strategy   = S
2. Culture   = C
3. Excellence = E

Therefore, the equation for excellence can be written as:

$$S + C = E$$

The importance of this equation in showing how strategy must be united with an organization's culture is fully explained by Hickman & Silva in the following quotation:

"Strategy and culture each contribute to the success of any organization. In the past, we have seen brilliant strategies bring great business success, and we have seen strong cultures survive great upheavals in the marketplace. In a few exceptional cases, a strong culture has overcome a stupid strategy, or a smart strategy has prevailed despite a weak culture, but don't count on such exceptions in our increasingly competitive and sophisticated business world. Long-term success and perennial corporate excellence require alloys of superior strategies and strong cultures" (Hickman & Silva, pp. 77–78).

Summary and Conclusions

This chapter outlines an approach that can be taken by institutions engaged in a quest for the creation and achievement of excellence in all programs and activities. The gist of this approach is that institutions must adopt major institutional strategies which are designed to (1) improve academic quality; (2) improve institutional management; and, (3) enhance and maintain fiscal stability. The four foundational strategies described in this paper are:

1. Encouraging strategic thinking throughout the university's organizational structure;
2. Building an institutional (corporate) culture;

3. Linking strategic planning with outcomes assessment; and
4. Implementing a model of institutional effectiveness (adoption of the 7-S Framework as a planning paradigm).

This chapter shows how college administrators can inculcate the principles derived from these strategies into every aspect of their mission, thereby, vastly improving institutional effectiveness and efficiency which in turn guarantees greater success in recruiting and retaining students.

References

Banta, T. W. (1985). Comprehensive program evaluation at the University of Tennessee, Knoxville: A response to changes in state funding policy. In T. Banta (Ed.), *Performance Funding in Higher Education: A Critical Analysis of Tennessee's Experience*. Boulder, CO: National Center for Higher Education Management Systems.

Banta, T. W. (1985). Uses of outcomes information at the University of Tennessee, Knoxville. In P. Ewell (Ed.), *Assessing Educational Outcomes. New Directions for Institutional Research, No. 47*. San Francisco: Jossey-Bass.

Banta, T. W. (1986). *The Instructional Evaluation Program at the University of Tennessee, Knoxville*. Unpublished paper, University of Tennessee.

Banta, T. W. and Fisher, H. S. (1986). Putting a premium on results. *Educational Record*, (Spring/Summer), pp. 54–58.

Hickman, C. R. and Silva, M. A. *Creating Excellence*. New York: New American Library, 1984.

Ouchi, W. G. *Theory Z*. New York: Avon, 1981.

Pascale, R. T. and Athos, A. G. *The Art of Japanese Management*. New York: Simon and Schuster, 1981.

Peters, T. J. and Waterman, R. H. *In Search of Excellence.* New York: Harper and Row, 1982.

Scofield, D. D. *The New Scofield Reference Bible.* "Luke 6:48". New York: Oxford University Press, Inc., 1967, p. 1087.

Waterman, R. H., Peters, T. J., and Phillips, J. R. Structure is not organization. *Business Horizons.* June 1980.

Chapter II

Cognitive Styles and Multicultural Populations

Dr. James A. Anderson
Associate Professor
Department of Psychology
Indiana University of Pennsylvania

In the past two decades, minority students have gained limited access to many predominantly white colleges and universities. Because minority students historically have been drawn from isolated rural areas or depressed residential areas where educational resources are sub-standard, most of these students are inadequately prepared to compete favorably at the university level against better educated, more affluent students. Even though the percentage of Blacks, for example, who complete high school has increased steadily since 1970, the percentage

Reprinted with permission from the *Journal of Teacher Education,* XXXIX (1), Jan/Feb, 1988 who go on to college has declined since 1980. And, among those who do attend, the retention rates reflect serious problems. Wilson (1984) suggests this decline will continue until 1990 unless there is an increase in the number of those attending universities and an increase in retention.

Through the early 1960s student retention received comparatively little attention in the professional literature. Even when it was discussed, it was largely an academic exercise but rarely resulted in program development. During the last few years a concern with retaining minority students has appeared in the academic literature and at institutions because minority students, especially Blacks, represent an important resource. For predominantly white universities and colleges, the increased retention of minorities would satisfy two needs: (a) it would allow the institutions to meet federal guidelines, and (b) it would allow the institutions to fill classrooms left vacant by the disappearance of the applicant pool that had been created by the legacy of the baby boom.

The concern with retention demands that institutions implement special programs to provide minority students with the kind of support necessary for their academic survival. At most institutions, retention programs are divided into two general categories: academic-sector programs and counseling programs. The amount of interaction between the two components varies from school to school.

Certain academicians (Astin, 1972; Demitroff, 1974; Beal and Noel, 1980) suggest that factors relating to academic potential are the major variables that can be usefully employed for predicting retention. There is considerable evidence that initial success in dealing with the academic environment of a college can be a crucial factor in facilitating academic survival (Pantages and Creedon, 1978). But, the academic approach to retention has focused on remediation rather than serving as a conduit into the general curriculum. Thus, students often succeed at remediation but do not achieve effectively beyond that.

By far the most commonly implemented retention programs come under the heading of counseling programs. These programs seek to intervene in the dropout process by facilitating student development of clear goals allowing the student to develop a more realistic self-perception, and provide a sort of early warning alarm when the student begins to nose-dive. A national examination of retention programs by the American College Testing Program reveals a preference for counseling programs among universities surveyed (Noel, Levitz, and Kaufmann, 1982). This is easy to understand because almost sixty percent of these programs were initiated by the Student Affairs sector, while fewer than a quarter (22%) were initiated by the academic sector.

One of the major assumptions of most retention programs is that the incoming student should adapt to all facets of the program. Adherents to this perspective assume that the standard retention model can effectively meet the needs (academic and emotional) of minority students and can match up with their values/attitudes concerning achievement. The lack of success among retention programs over the past decade suggests, that this assumption is incorrect.

Two critical factors underscore the minor success of retention programs. First, the programs themselves share the same model, one which emanates from basic educational learning theory and which reflects Anglo-

European notions about cognitive functioning, learning, and achievement. A second critical factor is that these programs have almost never attempted to identify the cognitive assets and learning preferences of non-white students. What has precluded the identification of the cognitive/learning styles of minority students in retention programs is the ethnocentric assumption on the part of whites that minorities do not have a valid and substantive cognitive framework which may be somewhat different but equally effective for them.

It is very easy to ignore the ethnocentric dominance of the theory and practice that exists in higher education. Universities and colleges are run (with a few exceptions) by white males, as are the departments in those same institutions. Departmental curriculums and retention programs, in fact the basic models of learning and achieving, derive from a very narrow white male perspective. The issue which has surfaced among many scholars who study minorities concerns the affirmation of the cognitive/learning styles among multicultural populations. Moreover, these same scholars call for an incorporation of knowledge about this cultural and racial diversity into university programs and ultimately into the teaching styles of individual instructors.

When students of color enroll at predominantly white colleges and universities, they are expected to adapt to the milieu of that environment. In fact, their capacity to adapt may significantly underscore their ability to achieve academically. But the question which is seldom addressed and incorporated into the success equation of these students is, "to what degree does their cultural and/or racial diversity impact their perception of this (new) learning environment?" Such a question is difficult to address when little attention has been given to the varied nature of cultural differences among multicultural populations.

Cultural differences, while sometimes highly apparent, can often be misleading. In much of the educational and social science literature the variable of "race" has been bandied about as the most important factor in accounting for group differences in various situations. This has occurred because race is often a very visible cultural difference; because other homogeneous characteristics can easily be grouped (rightly or wrongly) under it; and because it has been politically expedient to do so. Even when race is utilized as a critical variable, the validity of its conceptualization leaves much to be desired when applied to minority populations. For example, linguistic differences between Blacks and whites were initially conceptualized as nonstandard vs. standard, and then expanded to reflect more general cognitive differences (and in the case of Blacks, deficiencies). A similar example can be seen in psychology where race differences frequently appear in the perceptual skill of field dependence/independence. Minorities typically exhibit a preference for field dependence which is viewed as a more restrictive perceptual skill as are the other cognitive areas that correlate with it. Accurately conceptualizing racial and cultural differences is imperative so that superficial differences are not magnified, fluid differences are not treated as static, and our decisions about group behavior are based upon verifiable objectivity.

It does not take an incoming student of color long to realize that the university or college does not actually value cultural diversity in a practical sense. Ideally it may have been verbalized, but no mechanisms are in place to affirm the assets of the culturally diverse. Along the same lines, minorities in the general population recognize that this country is not a "melting pot" in which cultural diversity is respected and appreciated. Instead, it is a nation which is racially, ethnically, economically, geographically, culturally, and linguistically diverse. People of color are generally insulated and isolated away from the mainstream culture, such that only a minimal amount of acculturation, and even less assimilation, occurs. The greater the acculturation gap between an ethnic group and the mainstream culture, the more problems an ethnic group will experience. The problems, to a degree, reflect an inability by the persons of color to adapt to new demands and expectations, but the problems just as often stem from the tendency of majority group members to reject any cultural style which is not Anglo-European, or has not been decreed as valuable by them. Acculturation is such a critical factor that the variables of race and ethnic origin cannot be examined without considering its impact. In an examination of the interaction among these variables, some minority group members have become so acculturated that they identify with and behave like Anglo-European whites; while certain whites (southern Appalachian Anglos or Cajun Whites in Louisiana) who are compelled to endure economic and social constraints that prevent their penetration of the dominant middle-class, identify more closely with the cultural patterns of minority groups.

Towards a Conceptualization of Cultural and Cognitive Style

All components of a culture are built upon some basic conceptual system or philosophical world view, and the various cultural systems tend to include the same general themes (e.g., life, death, birth, morality, human nature, religion, etc.) Even though these beliefs appear across cultures, they can be viewed differently within each culture.

A conceptual system is a pattern of beliefs and values that defines a way of life and the world in which people act, judge, decide, and solve problems (Matthews, 1973). In each culture, reality is distinctively conceptualized in implicit and explicit premises and derivative generalizations which together form a coherent system.

The conceptual system becomes embedded in a particular network and is transmitted to its members through a complex matrix of socialization practices. The socialization process merely transmits choreographed patterns of behavior that a person learns to copy. However, Wilson (1971) suggests that something more important occurs: a person learns in a particular way. In other words, different cultures produce different learning styles. Wilson further suggests that cultural influence not only affects learning style, but the subtler aspects of perception and cognitive behavior as well. It would seem feasible that different ethnic groups with different cultural histories, different adaptive approaches to reality, and different socialization practices would differ concerning their respective cognitive/learning styles.

Both sociocultural and environmental factors are important in the development of any cultural style. Among the sociocultural factors are cultural values and beliefs, socialization practices such as child rearing and the system of rewards and punishment in the family, and sex role development. In terms of environmental factors, Witkin (1967) suggests that the degree to which the environment is variegated is important. In addition, the amount of environmental stress is crucial. Witkin explains that initially a child does not differentiate himself or herself from the environment, but as he or she matures, the process of psychological differentiation does occur and is affected by the above factors. The nature of this differentiation ultimately may be reflected in the child's cultural style.

Because the social, cultural, and environmental milieus of ethnic and racial groups differ, one should expect these differences to be reflected in their respective cultural/cognitive styles. Much of the literature in cross-cultural research supports this contention. Several researchers examined comparisons in cultural style for particular cognitive modes. For example, Allport and Pettigrew's (1957) classical study of the trapezoidal illusion illustrated African-European differences in children in their perception of movement. Bruner (1966) noted differences in perception of the conservative task between African and European children. Other studies have indicated differences between Blacks and Whites in linear perspective and depth perception (Killbride and Robbins, 1968) and in differential response to the semantic differential.

Work by Gagne and Gephart (1968) suggests that Blacks learn disjunctive concepts more easily than conjunctive ones, which is just the opposite from Europeans. Deregowski and Serpell (1974) found differences between African and European children in the way they classified pictures and objects. Ramirez and Price-Williams (1974a) showed that, in general, Mexican-American children tended to be more field-dependent while Anglo-American children were more likely to be field-independent. Other studies have identified ethnic differences on variables directly related to *patterns* of cognitive functioning. Ramirez andPrice-Williams (1971) compared the performance of Mexican-American and Anglo-American children on a series of tasks which involved telling stories about picture cards depicting educational scenes. Analysis of these stories revealed that those of the Mexican-American children were lengthier, indicative of greater verbal productivity, and included more characters than those related by Anglo-American children. Work by Lesser, Fifer, and Clark (1965) and Stodolsky and Lesser (1967) also indicated that different ethnic groups exhibit different patterns of intellectual performance, with each group achieving better in some areas than in others.

It should be noted that many of these studies utilize very disparate cultural groups as their samples: Africans and Mexicans vs. Anglo-Europeans and white Americans. But what about a case where different ethnic groups share the same country and to some degree a similar lifestyle? Is it possible that such groups may also manifest variant cultural styles?

The logical question arises concerning how American Blacks, for example, could manage to maintain a distinct cultural style while existing within the context of the dominant Western culture. Has not sufficient assimilation occurred so that one would expect a gradual erosion of non-adaptive stylistic behaviors?

Herskovitz (1958) contends that a significant number of African behaviors, values, and beliefs have been carried over by contemporary American Blacks. These survived because of the existence of institutions such as the Black church and the Black family. Probably a more significant explanation is that one's entire cultural style does not have to change because some behaviors are borrowed from the dominant culture. The cultural style of the sub-group (Blacks) has been sufficiently encapsulated within its own culture to prevent total assimilation.

Joseph White (1970) suggests that the psychological orientation to reality of Blacks is distinct from that of whites. Specifically, he feels that this orientation is more feeling-oriented or affective while that of whites is more objective or less affective. Although very broad, White's generalization serves as a basic working hypothesis upon which a more elaborate framework may be established concerning the possibility of differences in cultural or learning styles between the races.

Elaborating upon the same notion, Matthews (1973) contends that the major differences between the cognitive systems of Blacks and Whites is the degree to which the subjective, affective self is incorporated into the cognitive evaluation of reality. Matthews compellingly articulates:

> The Black cognitive process operates at the level of rationality and it comes out of an in-depth probing of reality . . . that this probing comes out of the affective track of a close-to-life experience and not out of the formulations or the postulates of a distant, depersonalized, dehydrated and unemotional logic. (p. 28).

He continues:

> But history, geography, environment and cultural conditioning operate to produce a different organization, structure, ranking and use of feeling and emotion in different peoples. This kind of difference affects the way in which the Black man perceives things, and this perception determines the Black man's organization, structuralization and presentation of his thought and vice versa. (p. 33)

As a result of this affective influence, Matthews suggests that the Black cognitive system can be referred to as the "aesthetic mind" or "feeling intelligence." The variant systems are simply extensions of each groups respective African and European heritage. Each culture has evolved around some world view within a philosophical system and it is this view that permeates the basic threads of cultural development.

The most characteristic feature of the African philosophical system is its focus on unity and connection; it is a view of extraordinary harmony. Humans are rhythmically united with nature and the universe. Humanity is a function of connectedness and interdependence of community, nature, and cosmos (Mbiti, 1970). All systems of thought and behavior, from the more formal sciences to simple practical concerns, are interwoven into a logical and functional system.

Thus the human being is physically, affectively, and cognitively united to the natural cycle of existence. Faith, reason, and the emotions are mutually dependent. Such a world view differs from the Europeans' in that things that are contemplated experienced, and lived are separable. For Africans there is no conflict between cognitive and emotional systems. This emphasis upon systematic unity appeared irreconcilable in European terms. Because it could not be accommodated to European systems of thought, the African way of thinking was considered nonlogical (Levy-Bruhl, 1966). This European approach to cognitive and emotional functioning had evolved into an institutional way of life in which certain clearly defined behaviors command a separation of affect and cognition.

Wober (1967) even goes so far as to suggest "that the sameness of cognitive style through all fields of behavior may not occur for Africans as it may for Americans" (p. 29). The illusion should not be created that Matthews

(1973) is referring to two separate and distinct cognitive systems, solely emotional for Blacks and solely unemotional for whites. On the contrary, both systems operate "at the level of rationality." This notion of an affective/cognitive style is not limited to Blacks, but appears in the conceptual systems of other groups of color.

Cultural Style and Classroom Learning

A pluralistic society is one which is populated by disparate ethnic groups and cultures. Social order is often a function of the tolerance level exhibited by the dominant group, and sub-groups are often indirectly stymied as they attempt to gain access to mainstream society. In American society the educational system is recognized as a primary vehicle to success-achievement. But, this system is built upon the type of European world view which was alluded to earlier and tends to benefit whites whose cultural style is more attuned to it. It seems feasible to suggest that the test-measured differences between Blacks and other groups of color and whites may be due to the fact that some minorities are not motivated towards optimal achievement within a Western educational setting. Cohen (1969) suggests that most school environments reflect a field-independent style, a style unfamiliar to Mexican-American, Puerto Rican, and Black children. Work by McNeil (1968) and McNeil and Phillips (1969) suggests that the school setting which is oriented towards whites does not parallel the learning system of Blacks. Even more important than the above suggestions would be an identification of the behavior patterns which possibly reflect differences in learning styles and which simultaneously contribute to differential academic performance.

Cohen (1969), Messick (1970), and Ramirez (1973) found that Mexican-American children tend to be field-dependent and do best on verbal tasks. They are also said to learn materials more easily which have humor, social content, and which are characterized by fantasy and humor; and are sensitive to the opinions of others. Anglo-American children do best on analytic tasks; learn material that is inanimate and impersonal more easily; and their performance is not greatly affected by the opinions of others. Kagan and Madsen (1971) studied the motivational styles of Mexican, Mexican-American, and Anglo-American children. They found that Anglo-American children were more competitive than either Mexican or Mexican-American children. The Mexican-American children also were more highly motivated in the cooperative setting than in the competitive. Ramirez and Price-Williams (1974a) also examined motivational style by utilizing the School Situation Picture Stories Technique (SSPST). They found that Anglo-American children scored higher on need achievement for self, in which the achiever is the primary beneficiary. In another study with the SSPST, Ramirez and Price-Williams (1974b) showed that Mexican-American children scored higher than Anglo-American children on need affiliation, and need succorance which refers to a need to be comforted.

In his book, *The Psychology of the Afro-American,* Adelbert Jenkins (1982) suggests that learning involves the application of one's meaningful framework to new situations in order to make sense of them. Learning becomes solidified when this mental reshaping becomes attached with some positive or negative affect. For example, one of the functions of language is the transmission of culture. The indications that language may be illustrative of cognitive style suggests that children internalize the cognitive style and perceptual views of their parents, along with the more overt aspects of cultural preference of the society. Cooper (1980a) suggests, for example, that as children hear and use holistic language features their preference for a holistic style is reinforced by most other contacts, both peer and adult, and if children are expected to behave in a particular way and in a particular situation where this behavior is reinforced, again, their preference for a particular style is reinforced.

In America, as white children leave the home and move on through the educational system and then into the work world, the development of cognitive and learning styles follows a linear, self-reinforcing course. Never are they asked to be bicultural, bidialectic, or bicognitive. On the other hand, for children of color, biculturality is not a free choice, but a prerequisite for successful participation and eventual success. Non-white children generally are expected to be bicultural, bidialectic, bicognitive; to measure their performance against a Euro-American yardstick; and to maintain the psychic energy to maintain this orientation. At the same time, they are being castigated whenever they attempt to express and validate their indigenous cultural and cognitive styles. Under such conditions cognitive conflict becomes the norm rather than the exception.

Cognitive Style and Cultural Groupings

Although the perception might be that cognitive styles are disparate entities, they actually operate along a continuum. Along this continuum, certain groups seem to cluster at one end or the other. There are many similarities in the world views and cognitive styles of certain groups of color that affect their fundamental perceptions of the world and how they choose to think about it and then interact with it. Table 1 categorizes some of the groups who share aspects of a world view; Table 2 identifies some of the cultural dimensions characteristic of each grouping.

As was alluded to earlier, the cognitive style of a group is strongly influenced by the cultural history of that group. There is no such thing as one style being "better than another," although in our country the Euro-American style is projected by most institutions as the one which is most valued. Cultural and cognitive conflict often occur when a group is asked to perform in a manner and setting which in some ways is foreign to their style or does not capitalize on it. In many critical areas of human functioning and behavior, the world view of the dominant group is indifferent to or conflicts with the world view of other groups in that culture (Jenkins, 1982; Greenspan, 1983; Baldwin, 1985; Day, 1985). Table 3 labels and compares various styles as they might impact classroom learning.

TABLE 1
Cultural Groupings of World Views

Non-Western	Western
• American-Indians	• Euro-Americans (primarily males)
• Mexican-Americans	• Minorities with high degree of acculturation
• African-Americans	
• Vietnamese-Americans	
• Puerto Rican-Americans	
• Chinese-Americans	
• Japanese-Americans	
• Many Euro-American females	

TABLE 2
Some Fundamental Dimensions of Non-Western vs.
Western World View

Non-Western	Western
• Emphasize group cooperation	• Emphasize individual competition
• Achievement as it reflects group	• Achievement for the individual
• Value harmony with nature	• Must master and control nature
• Time is relative	• Adhere to rigid time schedule
• Accept affective expression	• Limit affective expression
• Extended family	• Nuclear family
• Holistic thinking	• Dualistic thinking
• Religion permeates culture	• Religion distinct from other parts of culture
• Accept world views of other cultures	• Feel their world view is superior
• Socially oriented	• Task oriented

Multicultural Communication and Cognitive Style

According to Cooper (1981), a linguistics specialist at the University of the District of Columbia, almost no one has examined sufficiently the influence of cognitive style on the many aspects of speech (language) and writing. During the late 1960s and 1970s, sociolinguistic studies identified various non-standard dialects, particularly Black English, as valid language systems, and the validity of the dialects was generally accepted by the English profession. Grammatical and phonological aspects of dialect, however, cannot totally account for language differences among diverse populations. Because cognition and cognitive style directly influence the style and content of language (and vice-versa), it is imperative that researchers examine cultural differences from this perspective.

TABLE 3

Cognitive Style Comparison

Field-Dependent Relational/Holistic Affective	Field-Independent Analytic Non-Affective
Characteristics	**Characteristics**
1. Perceive elements as a part of a total picture.	1. Perceive elements as discrete from their background.
2. Do best on verbal tasks.	2. Do best on analytic tasks.
3. Learn material which has a human social content and which is characterized by fantasy and humor.	3. Learn material that is inanimate and impersonal more easily.
4. Performance influenced by authorizing figures expression of confidence or doubt.	4. Performance not greatly affected by the opinions of others.
5. Style conflicts with traditional school environment.	5. Style matches up with most school environments.

Speech patterns that are developed by a group equip them for effective interaction within their own community first, and then the larger society second. These speech patterns also reflect the cognitive style of that group. Ultimately, the speech patterns fuse with other aspects of communication for that group to provide reciprocal reinforcement for what is most valued by the group.

The white American or standard English communication style, is simply one of many. It was not the first and, objectively speaking, it is not the best. Simply, it is more functional for those with an Anglo-European frame of reference.

Generally speaking, nonwhite and rural white students with a minimal degree of acculturation exhibit a communication style that is at variance with the Western communication style. Table 4 compares some of the key features of speaking/writing that differentiate holistic vs. analytical thinkers. These features have been identified by several researchers (Cooper, 1980b; Vygotsky, 1978; Cohen, 1969).

Another aspect of the holistic (nonwestern) communication style that distinguishes it from the analytical involves the degree to which symbolic and concrete imagery are utilized as a communicative tool. For some ethnic groups the extended use of imagery represents one of the dominant ways of thinking, conceptualizing, writing, and speaking. This process seems to be more characteristic of those groups that emphasize the affective, holistic, relational style.

According to Matthews (1977), the concrete symbolic image is the tool of a cognitive style preferred by Blacks (and other people of color) throughout the world. He states:

> In Black use the thought is generated through the use of a picture concept (Visualization) rather than through the use of . . . theoretical statement . . . a picture of the thing as it really exists is put before the mind and imagination . . . one proceeds through visual as against non-visual thinking. (p. 16)

Many cultures utilize symbolism, but European symbolism is empirical and abstract while that of holistic cultures is concrete and drawn from everyday life. Moreover, the latter communicates the affective aspects of that life. The Anglo-European style *does* often make use of symbolic imagery but selectively categorizes its usage under headings such as "poetry" and "literature." Table

TABLE 4

Comparison of Features in the Writing Styles of
Holistic vs. Analytical Thinkers

Holistic (Non-Western)

Holistic (Non-Western)	Analytical (Western)
1. Descriptive abstraction	
2. Word meaning based on content	1. Analytic abstraction
3. Use few synonyms	2. Formal meanings for words
4. Use few comparisons	3. Use many synonyms
5. Use relational and institutional classification	4. Use many forms of generalization comparison
6. Tends to use second person "you," reflects group identity, tends to pull reader in as part of the writing	5. Use hierarchical modes of classification
	6. Can easily adopt a third person viewpoint in writing and speaking, is objective, reflects separate identity from what is going on

TABLE 5

Comparison of Form and Function of Symbolic Imagery Between
Disparate Cognitive Processes (Speech & Writing)

Non-Western

Non-Western	Western
1. Visual (pictorial thinking)	
2. Thought is perceives as . . . living thing, wholistic thing, doing thing	1. Notions or theoretical statements
3. Imagery is intensely affective with cultural base	2. Thought is . . . Mentalisitc, devitalized, static
4. Extensive expression of concrete emotional words and heightened use of metaphors	3. Imagery minimizes affective associations
5. Medium is the message	4. De-emphasis on such unless in specialized disciplines or situation
6. Medium motivates and socializes	5. Medium communicate the message
7. Introduces self into objective analysis of events	6. Things must be contemplated before they motivate
	7. Removes self

5 represents a comparison of the stylistic difference in the utilization of symbolic imagery.

One of the most critical problems encountered by students of color is that secondary school teachers and college faculty are not equipped to identify, interpret, and respond to the variant styles of multicultural populations. A communication gap exists between their teaching style and the students' indigenous learning styles. Thus, the symbolic, affective, reality-based approach to learning of some students will not only be misconstrued but also branded as deficient. For example, the writing/speaking styles of Mexican-American, Black-American, and Puerto Rican-American students are frequently viewed as "too flowery," too subjective, involving an excessive use of metaphors, utilizing the wrong tense of verbs, etc. What is a valuable and valid communication process under one cognitive style becomes a deformed example of cognitive/linguistic deficits under another.

Even more salient are the problems that students of color encounter when they attempt to adapt their styles to the theoretical, often abstract, reasoning process utilized in mathematics and the hard sciences. Most courses in both areas utilize a format in which the teaching of the theory (in an abstract sense) precedes any practical application or direct experience (like laboratory exercises). The implicit assumption is that this is the proper sequence of training because, historically speaking, this has always been the approach to mathematics and science instruction; and because this approach coincides with the Anglo-European cognitive style, especially that of males. The opposite approach, in which direct experience precedes discussion of formal concepts and laws, is not as valued and, hence, utilized as much by teachers. McDermott, Piternick, and Rosenquist (1980) found such an approach to be extremely successful in the development of a curriculum in physics and biology at the University of Washington. Brown (1986) at the University of San Diego discovered that the same approach worked in her mathematics lab. Successful programs which serve minority populations utilize other dimensions of the holistic/affective. For example, one of the most successful science/pre-med programs of the last decade has been the SOAR program at Xavier University in New Orleans. The program builds confidence and skill in its predominantly black population by creating an aura of family in which cooperation is highly valued, bonding between the students and faculty is encouraged, and a maintenance of positive ethnic identity is fostered. Learning occurs in a socially reinforcing environment. Incidentally, the director of the program is a white male. One does not have to be the same race/ethnicity to identify and capitalize upon the cultural/cognitive assets of minority populations.

Why is it that teachers refuse to alter their teaching styles when student performance and classroom behavior suggest that communication problems exist? Sandefur (1987) suggests that many students are shortchanged because most mathematicians refuse to address the issue of the practical applications of mathematics before the theory; as a result, many students lose interest early. He then cites several reasons why this is the case, and these reasons can be generalized to the rigidity of teaching styles in most disciplines. First of all, many teachers are enamored with the abstractions of their discipline. To alter their approach would be to "defile their art." Another reason is that they enjoy the mystique that surrounds their art. They represent the elite who have mastered the understanding of theory and application. A third, and probably most important reason, involves the interaction between laziness and fear. Instructors often become rigid in their teaching styles and are too lazy to change. They have become stagnant, and anything that suggests a change in the status quo produces an undercurrent of anxiety. Confronted with an increase in multiethnic enrollment in their courses or the university in general, they retreat even further into their stagnant ways and bristle at the thought of change.

Conclusion

At the superficial level, cross-racial, cross-ethnic teaching, learning, and/or service delivery occurs when the persons interacting are of different racial or ethnic identities. When one adds to this equation the differences in degree of acculturation and type of cognitive/learning style, the examination and explanation of these differences becomes more complicated and the urgency to identify the critical dimensions associated with them more pronounced. Whereas it was once fashionable and sometimes academically rewarding to deny the existence of cultural assets and variations among non-white populations, social scientists and researchers now recognize that such traditional approaches have become anachronistic. The failure of retention programs, the ineffectiveness of service delivery to multiethnic populations, and the inability to produce effective communication between majority and minority members are glaring examples that new models and approaches must evolve which not only deal with debilitating misconceptions about minorities, but, more important, also operate within a framework of equal respect and appreciation for the similarities and differences among groups.

Boyer (1983) offers ten critical dimensions of cross-racial, cross-ethnic teaching, learning, and service delivery. Two of these are very apropos concerning the issue of disparate cultural/cognitive styles. One dimension addresses the importance of altering the belief systems of educators and professionals that historically have supported academic racism. This can be done in a variety of ways but, at a minimum, must be incorporated into the inservice training and professional preparation of educators. A second relevant dimension concerns how to affect the experiential base of persons (educators and professionals) who operate by rigid unidirectional styles: In other words, how does one offset the impact of negative affective responses to culturally diverse student behavior?

A different set of understandings about the way diverse populations communicate, behave, and think needs to be developed by educators. Until this occurs, educators will continue to stagnate in the dark ages and educators will provide lip service rather than action to the egalitarian values associated with pluralism and multi-culturalism. Minority researchers have begun to affirm the value and viability of their cultural strengths and will expect institutions to reciprocate. As professional educators, we must settle for nothing less.

References

Allport, G., & Pettigrew, T. (1957). Cultural influences on the perception of movement: The trapezoidal illusion among Zulus. *Journal of Abnormal Social Psychology, 55* (1), 104–113.

Astin, A. W. (1972). College Dropouts: A national profile. *ACE Research Reports, 1,* 1–71.

Baldwin, J. A. (1985). Psychological aspects of European cosmology in American society. *Western Journal of Black Studies, 9* (4), 216–223.

Beal, E. P., & Noel, L. (1980). *What works in student retention.* Iowa City, IA: The ACT National Center for Higher Education Management Systems.

Boyer, J. (1983). The ten most critical dimensions of cross-racial, cross-ethnic teaching and learning. *Educational Considerations, 10* (3), 2–4.

Brown, M. (1986, November). *Calculus by the dozen: A retention program for undergraduate minority students in mathematics based majors.* A paper presented at the second annual Conference on Black Student Retention, Atlanta, GA.

Bruner, J. et al. (1966). *Studies in cognitive growth.* New York: John Wiley.

Cohen, R. A. (1969). Conceptual styles, culture conflict and nonverbal tests of intelligence. *American Anthropologist, 71,* 828–856.

Cooper, G. (1980a). Different ways of thinking. *Minority Education, 2* (5), 1–4.

Cooper, G. (1980b). Everyone does not think alike. *English Journal, 60,* 45–50.

Cooper, G. (1981). Black language and holistic cognitive style. *Western Journal of Black Studies, 5,* 201–207.

Demitroff, J. E. (1974). Student persistence. *College and University, 47,* 553–567.

Deregowski, J. B., & Serpell, R. (1974). Performance on a sorting task with various modes of representation; A cross-cultural experiment. In M. Cole & S. Schribner (Eds.), *Culture and thought: A psychological introduction.* New York: John Wiley.

Day, M. W. (1985). *The socio-cultural dimensions of mental health.* NY: Vantage.

Gagne, R. M., & Gephart, W. (1968). *Learning research and school subjects.* Itasen, IL: F. E. Peacock.

Greenspan, M. (1983). *A new approach to women and therapy.* New York: McGraw-Hill.

Herskovitz, M. (1958). *The myth of the Negro past.* New York: Harper.

Jenkins, A. H. (1982). *The psychology of the Afro-American*. Elmsford, NY: Permagon Press.

Kagan, S., & Madsen, M. C. (1971). Cooperation and competition of Mexican, Mexican-American, and Anglo-American children of two ages under four instruction sets. *Developmental Psychology, 5*, 32–39.

Killbride, P., & Robbins, M. (1968). Linear perspective, pictorial depth perception and education among the baganda. *Perceptual and Motor Skills, 21* (2), 161.

Lesser, G. S., Fifet, G., & Clark, D. H. (1965). Abilities of children from different social class and cultural groups. *Monographs of the Society for Research in Child Development, 30* (4).

Levy-Bruhl, L. (1966). *How natives think*. New York: Washington Square Press.

Matthews, B. (1973). *Black cognitive process*. Unpublished paper, Howard University, School of Social Work, Washington, DC.

Matthews, B. (1977). Voice of Africa in the Diaspora. *New Directions, 4*, 16–19.

Mbiti, J. (1970). *African religions and philosophies*. Garden City, NY: Doubleday.

McDermott, L. C., Piternick, L. K., & Rosenquist, M. L. (1980). Helping minority students succeed in science. *Journal of College Student Teaching, 10* (1), 135–140.

McNeil, K. (1968). Semantic space as an indicator of socialization. *Journal of Educational Psychology, 59*, 325–327.

McNeil, K., & Phillips, B. (1969). Scholastic nature of responses to the environment in selected subcultures. *Journal of Educational Psychology, 60* (2), 79–85.

Messick, S. (1970). The criterion problem in the evaluation of instruction: Assessing possible, not just intended outcomes. In M. C. Wittrock & G. D. E. Wiley (Eds.), *The evaluation of instruction: Issues and problems*. New York: Holt.

Noel, L., Levitz, R., & Kaufmann, J. (1982). *Organizing the campus for retention*. Iowa City, IA: ACT National Center for the Advancement of Educational Practices.

Pantages, T. J., & Creedon, C. F. (1978). Studies of college attrition: 1950–1975. *Review of Educational Research, 48*, 49–101.

Ramirez, M. (1973). Cognitive styles and cultural democracy in education. *Social Science Quarterly, 53,* 895–904.

Ramirez, M., & Price-Williams, D. R. (1971). *The relationship of culture to educational attainment.* Houston, TX: Rice University, Center for Research in Social Change and Economic Development.

Ramirez, M., & Price-Williams, D. R. (1974a). Achievement motivation in Mexican-American children. Unpublished manuscript. Houston, TX: Rice University.

Ramirez, M., & Price-Williams, D. R. (1974b). Cognitive styles of children of three ethnic groups in the United States. *Journal of Cross-Cultural Psychology, 5* (2), 212–219.

Sandefeur, J. (1987, January 21). Mathematics teachers are too lazy to change their ways: As a result, teaching is stagnant. *Chronicle of Higher Education,* pp. 40–41.

Stodolsky, S. S., & Lesser, G. S. (1967). Learning patterns in the disadvantaged. *Harvard Educational Review, 37* (4), 546–594.

Vygotsky, L. S. (1978). *Mind in society: The development of higher psychological process.* Cambridge, MA: Harvard University Press.

White, J. (1970). Guidelines for Black psychologists. *The Black Scholar, 1* (5), 52–57.

Wilson, R. (1971). *A comparison of learning styles in African tribal growth with Afro-American learning situations and the channels of cultural connection: An analysis of documentary material.* Unpublished doctoral dissertation, Wayne State University, Detroit, MI.

Wilson, R. (1984, February). *Minority underrepresentation in post-secondary education.* Paper presented at the Illinois Committee on Black Concerns in Higher Education Workshop, Normal, IL.

Witkin, H. A. (1967). A cognitive style approach to cross-cultural research international. *Journal of Psychology, 2* (4), 233–250.

Wober, M. (1967). Adopting Witkin's field-independence theory to accommodate new information from Africa. *British Journal of Psychology, 58* (1–2), 29–38.

Chapter III

Project Far: A Blueprint for College Student Retention

Ms. Rebecca T. Gates
Former Director, Project FAR
Delaware State College

Introduction

The recent trend of declining enrollment and lowered academic achievement continues to pose a challenge for many colleges and universities across the country. The historically black colleges and universities have begun to address the problem by forming Retention Task Forces and/or implementing remedial programs for high risk students. Too often the results have been discouraging. However, Delaware State College's approach to the problem through its Project Freshman Attrition Reduction (FAR) Program has had very positive results.

Project FAR had its beginnings at Delaware State College in 1976 in response to a student attrition rate of

41 percent and a probation rate of 56.5 percent. Dr. Randell Trawick, former Director of Counseling, and David Reynard, Director of Institutional Research, conceptualized and implemented the Attrition Reduction Program (ARP) utilizing an affective approach to reduce the number of dropouts and improve the academic standing of students, particularly at the freshman level.

Through the program's curriculum and intervention strategies, the attrition rate declined. The dropout rate went from 44 percent to 26 percent. In addition, the number of entering freshmen who graduated four years later rose 22 percent and the probation rate went from 56.5 percent to 31.9 percent. Specifically, the project focuses on development of self concept, educational values, and study attitudes as they relate to academic achievement. The major goals are to create student awareness of both cognitive and affective factors contributing to attrition and to offer services to facilitate adjustment to college. To achieve the goals of the program, three components were identified and implemented. These components are defined and discussed below.

Program Components

The schematic representation pictured below (figure 1) shows how the three components Preventive, Early Warning and Rehabilitative work together to produce program outcomes. The *Preventive Component* consists of a one-credit weekly orientation class required for all freshmen. The *Early Warning System* is the administration of Dr. Alexander Astin's dropout predictor instrument to all freshmen to target those students with a higher than average probability of dropping out. The

Rehabilitative Component provides counseling, tutoring, human development workshops and dormitory counseling program. These components work together to identify target students early in the academic term and provide services to the students. The program outcomes are: (1) improved self concept, (b) decreased within semester withdrawals, (c) reduction in between semester attrition, (d) improved grades, and (e) reduction in probation rates.

Evidence to Support Claims of Reduction in Attrition

Total attrition was analyzed for short and long-range effects. In the short run, freshmen attrition rates were dramatically reduced. In the baseline year 1975–76, 41 percent of freshmen who entered in the fall of 1975 did not return in the fall of 1976. During the period of the program from 1977–1981, the total freshmen attrition rate was reduced to 16 percent. These data also compare favorably with the 46 percent freshmen attrition rates reported nationally by colleges with open and modified admissions policies.

The long-term effectiveness of the program is supported by a significant decrease in the percent of college-wide, within semester withdrawals between 1976 (baseline year—no program) and 1979 when the program had been operational for three years $(X^2(1) = 5.02, p.05)$. See Table 1.

Evidence of Reduction in Probation Rates

Academic probation is often a precursor of college attrition and many aspects of the program were designed

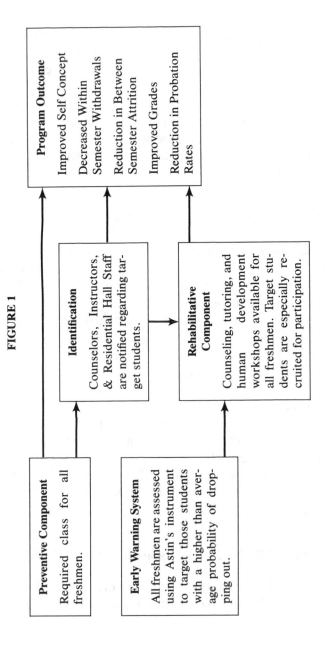

FIGURE 1

TABLE 1

Relative Frequency of Within Semester, College-Wide Withdrawals
(n = 8475)

Semester	Total Within Semester Withdrawals College-Wide	Total College Enrollment	% of Entire Student Population
Fall 1976 (baseline)	259	2167	12.0
Fall 1977	225	2128	10.6
Fall 1978	204	2129	9.6
Fall 1979	200	2051	9.8

TABLE 2

Freshmen Probation Rates from 1976 through 1979
(n = 1726)

Semester	Total Freshmen Enrollment	Freshmen on Probation (GPA less than 1.7)	% of Freshmen Population
Fall 1976 (baseline)	458	232	56.5
Fall 1977	422	118	28.1
Fall 1978	467	128	27.4
Fall 1979	379	121	31.9

to improve academic standing. Therefore, the short-term effects of the program were evaluated by comparing freshmen probation rates (GPA less than 1.7/4) for the baseline year with the three subsequent years during which the program was in operation. See Table 2.

In 1981, the Project came to national attention and was validated as an exemplary program by the U.S. Department of Education. In 1982, the program was funded by the National Diffusion Network under the U.S.D.E. umbrella to serve as a developer demonstrator providing materials and technical assistance to colleges and universities wishing to implement a similar program. During the past four years, forty colleges have adopted Project FAR. Of the forty adoptees, several have been historically black institutions. Included in this group are Bowie State College, Cheyney University and Lincoln University. In addition, the following historically black colleges have also received Project FAR training: Bishop College, Morgan State University, Savannah State College and Tennessee State University.

For additional information about Project FAR, please contact the Director, Project FAR, Counseling Department, Delaware State College, Dover, DE 19901

Chapter IV

Teaching Problem Solving in General Chemistry at a Minority Institution*

J. W. Carmichael, Jr., Professor of Chemistry and Premedical Advisor; Sr. Joanne Bauer, Instructor of Chemistry; and Donald Robinson, Professor of Chemistry, Xavier University of Louisiana

Many students have difficulty learning science because of poor problem-solving skills. This is particularly common in entry-level courses. Programs intended to help students who perform poorly in general chemistry have usually focused on additional exposure to chemistry

*The majority of the problem-solving activities described herein were developed or refined with support from the Josiah Macy, Jr., Foundation and the Health Careers Opportunity Program (Division of Disadvantaged Assistance, Bureau of Health Professions, HRSA)

content through preparatory courses and tutoring. While this approach is effective for some students, recent studies [5–9] suggest that more of them would benefit from increased attention to development of problem-solving skills, even if this left less time for content. We are attempting to integrate the teaching of chemistry content with two approaches to improving analytical thinking skills.

The general chemisty course at Xavier University of Louisiana (a small, predominantly black institution in New Orleans) is designed to serve the needs of the educationally disadvantaged student while still providing a challenge for the well-prepared student. The course has been "standardized" for more than 10 years, with the chemistry department as a whole, not individual teachers or texts, determining content and pacing in all sections of the course. In addition, there has been a conscious on-going effort to develop a comprehensive educational delivery system. The course has constantly evolved as various ideas have been tried, evaluated, modified, or discarded. Previous phases of development have been outlines in earlier reports [1,4,10]. The current form emphasizes some of the most unconventional activities—those designed to improve problem-solving skills.

An Overview of Course Structure and Materials

General chemistry at Xavier is a two-semester sequence with separate lecture and laboratory courses. In order to gain three semester hours credit for learning assistants grade the posttest. Each trio submits a single answer sheet. It is immediately graded, yielding one point for either all or none of the group.

During the remaining 15 minutes, selected trios explain their answers to these "problem-solving exercises." Both correct and incorrect solutions are discussed so that students not only learn how to solve the problem, but can analyze where they went astray.

Working in assigned trios is an idea acquired from Tom Taylor [11]. The approach used in solving and discussing problems is an adaptation of the "Whimbey method" of teaching problem solving [13]. In brief, the approach assumes that the type of problem solving one needs for most exams (breaking a complex problem into steps, each of which is solved sequentially) is a skill similar to hitting a baseball and can be taught in the same way. First you demonstrate how to hit/solve; then you allow the learner to practice while you are monitoring the process and making suggestions for improvement.

The critical point in teaching problem solving is to get both teacher and student to ignore our cultural bias that problem solving done rapidly or in one's head is "good." Instead, we focus on careful, methodical work in which all steps are clearly articulated and constantly checked for accuracy. Placing students into problem solving/monitor pairs (trios in our case) is one mechanism Whimbey suggested for implementing this procedure in a classroom.

Third 45-Minute Period

If all members of a trio pass the first version of the posttest (make a 90 percent grade or better), they are given 20 points for the posttest and one "group point" because all of the trio made 80 percent or better and are dismissed from the class for the day.

This leaves the learning assistants with the trios in which one or more students did not obtain the 90 percent grade. These students are given back their "incomplete" papers. The members of the trio who passed act as tutors, teaching both chemistry and thinking skills with supervision and help from the learning assistants.

At this point the practice problems used are those in the handbook associated with the question(s) that the individual student missed. The approach is again that of Whimbey and Lochhead [13]. The instructor and learning assistants circulate in the classroom, making certain that tutoring is being done properly and serving as tutors themselves when needed. Trio members who passed the first posttest are motivated to participate by the "group point" that they will receive only if those they are tutoring score 80 percent or better on the test to come.

Final 45-Minute Period

Students who did not pass the first posttest are given a different version and asked to work only problems equivalent to the ones they did not pass before. As each student finishes, the quiz is graded by a learning assistant who pays particular attention to detail, requires the student to explain all steps in answering questions, and asks the student to solve additional problems when he or she does not understand sufficiently. Students who take the second quiz are given two points for each of the questions they worked correctly on either attempt (for a maximum of 20 points total). All trios whose members sored 80 percent or better after the second testing receive an additional "group point."

Drill Characteristics

The general chemistry drills at Xavier have a number of important features. First, much of the stigma associated with remediation is eliminated. All students go through the same procedure, gaining exemption from class through good performance rather than being assigned extra work because of poor performance.

Second, the concept of trios is used to promote development of a peer support system centered around academics. Students in the same trio frequently sit together in other classes and meet to study together outside of class. Third, the student/teacher ratio decreases with time, thus allowing more personal attention after specific problem areas are identified. Fourth, tutoring of chemistry content is integrated into the course itself.

Finally, the overall approach is significant. There is a constant and overriding emphasis on accuracy, attention to detail, and rechecking everything. Vocalization plays a major role in the process. Thus the tutors, whether peers within a trio or learning assistants, teach both chemistry content and thinking skills. The operating procedure is best described by its developer, Arthur Whimbey [4]:

They do not just give answers or demonstrate the way a problem can be solved. Instead, they help students find answers for themselves. This may mean guiding a student in careful reading of a section of the text or asking a probing question which suggests the way a problem can be solved or a conclusion can be reached from knowledge available to the student. In addition, students are frequently asked to explain all of the steps in reaching an answer, especially if the answer is wrong, but also whenever the tutor suspects that a student's understanding of a topic is inadequate. In demonstrating how a particular type of problem is solved, the tutor may instruct the student and provide explanations, but he or she requires the student to carry out the actual activities and computations. The tutor never simply works a problem, leaving the student to watch. After helping a student to solve a problem, the tutor then asks the student to solve one or more additional problems and to explain the steps as he or she proceeds. Through tutoring, the tutor constantly watches for instances where a student arrives at an answer without sufficient concern for accuracy, reasons, or necessary cognitive steps. The tutors understand that their goal is both to teach chemistry and to eliminate the student's need for tutoring.

Students enrolled in the course and learning assistants are trained to use the approach during the first drill session of the semester when there are no previous lectures on which to quiz. The process of training is facilitated by the fact that approximately half of the students enrolling in Xavier's general chemistry sequence have previously completed a special bridge summer program, Project SOAR (see box), which also uses the same approach to systematic, logical problem solving.

"Doing Science" in the Laboratory

Drill sessions in general chemistry at Xavier emphasize the type of logical, sequential problem solving needed to comprehend texts, answer class exam questions, and score well on standardized tests such as the SAT, GRE, and MCAT. The laboratory courses are also designed to promote intellectual development, but with a different focus.

Our general chemistry laboratory program is organized to mirror the actual development of science as much as possible. To accomplish this, we require the student to use the type of analytical reasoning used in such development. This means reversing the usual relationship of lecture and laboratory. Our laboratory *precedes* lecture to provide a sequence of hands-on experience, data collection, analysis, idea testing, and formalizing those ideas.

The emphasis is on collecting data and analyzing it in a logically consistent and complete manner, rather than obtaining correct scientific facts. It is therefore possible to obtain an "A" for a laboratory report with a detailed, thorough analysis, even though the conclusion reached is not scientifically correct. Such an approach encourages students to develop the type of reasoning really needed to carry out a scientific project. They do not work toward some predetermined answer, as is all too often the case in the usual chemistry laboratory experiment.

We continue the process in lecture by introducing topics wherever possible via data tables and laboratory analyses. Thus, for example, the behavior of gases would typically be introduced in the laboratory. The students collect data (including some that is purposely extraneous) from a sample of gas, look for qualitative relationships among the variables, further refine the relationship by determining what (if any) equation describes the data, and then apply the results in some manner.

The following lecture presentations would begin with similar sets of data and analyses, proceeding through a set of system-specific equations to culminate in

$PV = nRT$. Laboratory emphasis is on the thinking process used to obtain the answer, with less regard to the scientific validity of the answer obtained. Any misconceptions thus obtained would be corrected in lecture, through analyzing error-free data.

The primary basis for our laboratory program is the *Laboratory Manual*, which contains a series of Karplus-Piagetian-oriented learning cycles [6]. The overall program has been discussed in more detail elsewhere [10].

The Bottom Line—Does It Work?

This modification of teaching techniques, including the emphasis on problem solving, is designed to increase success in chemistry for all students. However, because XU is a historically black institution, the overall goal of the program is to increase the number of black Americans who are prepared to enter science-related careers. The success of the program is reflected by the following:

• More than two-thirds of the students who enter general chemistry at XU now complete the year-long sequence with a "C" or better, versus one-third before modification.

• Performance on the American Chemical Society's *Cooperative Examination in General Chemistry* has doubled and is now at the national average.

• Students in general chemistry have consistently rated the course highly in a university-wide course evaluation procedure. In addition, a departmentally-administered questionnaire indicated that more than 80 percent of the students *liked* general chemistry, even though they felt they worked harder in the present course than they would have in the traditional one. The feeling that they had to work harder was compensated by the fact they also felt they learned more than they would have in a traditional course.

• Finally, modifications in general chemistry have played a major role in Xavier's dramatic growth in placement of students into science-related graduate or professional schools. In the past 10 years, the university has doubled the number of students who enter such fields yearly. For example, in 1984–85 Xavier, an institution with a full-time undergraduate enrollment of less than 1,600, was first in the nation in placing black Americans into colleges of pharmacy and second nationally in placing blacks into medical school.

Conclusion

While committed to all students, Xavier University has a special interest in providing remediation for the underprepared, since they as well as their better-prepared counterparts are needed if underrepresentation of blacks in science-related fields is to be corrected. The emphasis on problem solving in general chemistry is one component of an overall effort in the sciences and at the university as a whole to assist students to reach expected levels of performance. Although developed primarily to assist minority students, the success of XU's problem-solving approach has obvious application at other institutions interested in serving significant numbers of underprepared students in science courses.

Acknowledgment

Special note should be given to Arthur Whimbey, who devoted considerable time helping to implement the problem-solving approach in the recitation sessions.

References

1. Carmichael, I. W., Jr. "General Chemistry by PSI at a Minority Institution." *Journal of Chemical Education* 53:791–92; 1976.

2. ———. "Improving Problem-Solving Skills: Minotiry Students and the Health Professions, Summer Programs for Underprepared Freshmen." *New Directions for Learning Assistance* 10:37–45; 1982.

3. ———, J. Hunter, *et al.* "Project SOAR: Stress on Analytical Reasoning." *The American Biology Teacher* 42:169–73; 1980.

4. ———, M. A. Ryan, and A. Whimbey. "Cognitive Skills Oriented PSI in General Chemistry." *Journal of Developmental and Remedial Education* 3:4–5; 1979.

5. Collea, F. P., and S. G. Numedal. "Development of Reasoning in Science (DORIS): A Course in Abstract Thinking." *Journal of college Science Teaching* 10:100–04; 1981.

6. Karplus, R. "Science Teaching and the Development of Reasoning." *Journal of Research in Science Teaching* 14:169–72; 1977.

7. Lawson, A. E., and W. Wollman. "Using Chemistry Problems to Provoke Self-Regulation." *Journal of Chemical Education* 54:41–43; 1977.

8. McDermott, L. C., L. K. Piternick, and M. L. Rosenquist. "Helping Minorities Succeed in Science I–III.'" *Journal of College Science Teaching* 9:135–40, 201–05, 261–65; 1980.

9. Renner, J. W., and A. E. Lawson. "Promoting Intellectual Development Through Science Teaching." *The Physics Teacher* 11:273–76; 1973.

10. Ryan, M. A., D. Robinson, and J. W. Carmichael, Jr. "A Piagetian-Based General Chemistry Laboratory Program for Science Majors." *Journal of Chemical Education* 57:642–45; 1980.

11. Taylor, T. E. "Maximizing Success Through Peer Teaching." Presentation at the 35th Southwest Regional ACS Meeting, Austin, TX, 1979.

12. Whimbey, A., J. W. Carmichael, Jr., *et al.* "Teaching Critical Reading and Analytical Reasoning in Project SOAR." *Journal of Reading* 24:5–10, 1980.

13. ———, and J. Lochhead. *Problem Solving and Comprehension: A Short Course in Analytical Reasoning*. Philadelphia: Franklin Institute Press, 1979.

Chapter V

Models of Community Resources for the Enhancement of Black Student Retention

Dr. David W. Hoard
Program Manager
Academic Affairs
The Washington Center

Local towns and communities have traditionally played a crucial role in the success of historically black colleges and universities. The perception of community among the college, the locality and the black church is th cornerstone of black education. This relationship has brought financial, emotional, intellectual and spiritual help in a reciprocal fashion to everyone involved.

During the 1960s and 70s, more educational and professional opportunities developed for the black population and strained the traditional relationships between the community, churches and historically black colleges. No longer did black students automatically enroll in their nearby historically black institution. Predominantly white colleges expanded their traditional base to include people of other ethnic and cultural diversities. This was due to legal ramifications as well as financial necessity. As a result, student enrollments severely began to decrease within the historically black colleges and universities to a point where today some institutions are fighting for their existence.

In *Blacks in College,* Dr. Fleming comments on the strong traditional base between the community, the church and black colleges. She suggests that black students continue to receive a better and much broader education within a traditionally historic black college despite the perceived advantages offered by white institutions. Despite the larger financial advantages; larger libraries, better laboratory facilities, more faculty and more accommodating study facilities, white colleges cannot provide the breadth of role models and mentors that inspire black students to higher levels of achievement. Due to low minority student enrollment, white institutions cannot provide diverse positive support systems. The lack of such support systems manifests itself negatively in peer relationships. Moreover, the lack of strong friendship bonds negatively impacts student performances particularly when competition is a variable. Support plays a vital role in achievement by black students at historically black institutions. Without a viable support network student achievement will suffer and the problem of student retention will be exacerbated.

This thesis is further suported by Dr. Fleming in her discussion of role models, mentors and strong community ties as important features for the survival of minority students today. This paper presents proven and successful support systems developed and used by The Washington Center, a nonprofit experiential education institution based in Washington, D.C., for retention of its minority students who participate in internship and symposia programs. The support systems include monitoring, symposia, speaker series, adopt a church and more. These support mechanisms utilize little or no funding considerations and could be used on various campuses throughout the country. Time and effort are all that is necessary to increase student retention potential at many of today's institutions.

Community based support systems established by The Washington Center have proven to be of value to students, the Center and the community itself. The Washington Center's academic programs provide structured, semester long internship and short term issue-oriented symposia to undergraduate students from colleges and universities around the country. Via the internships, students are placed in congressional and executive branch offices, business and nonprofit organizations in Washington, D.C. In addition, the academic programs expose the students to important national issues and policy makers through speaker programs and academic seminars. The Center also concentrates on exposing the students to the community-based efforts of Washington.

The Washington Center provides support systems such as mentoring for minority students that ease their transition into the new and unfamiliar realms of the Washington area as well as the professional workplace. Not only have these support systems helped with the Center's retention of ethnic minority students, but the programs have aided the students through methods of networking and other kinds of support systems that can be utilized once they have returned to their educational institutions by making use of their newly discovered and refined skills and self confidence to complete their education.

The community based support systems have been crucial in terms of the Center's minority students participants and their development. These systems can be implemented on campuses to educationally stimulate college students to strive toward success. As they develop more self confidence, students also build upon their leadership skills. Finally, students have been able to develop resources to ensure financial support for their desires to continue their education and pursue graduate studies. Overall, The Washington Center has demonstrated its systems and has provided students with a light at the end of the tunnel to hopefully ignite their interests in fulfilling their educational pursuit within their original institution.

Methodology

The methodology used in completing this paper included:

1. A review of subject-related literature and relevant data on minority student retention rates and graduate study statistics.
2. A review of records on all minority students enrolled in The Washington Center internship programs since 1975. Such records included demographic and academic majors, internship placements, original internship learning goals, midterms and final evaluations.
3. Data were collected through a questionnaire on program impact, which will be used to survey all

minority program participants as well as a random sample of white participants since 1975.

4. Planned case studies were developed on data from the above sources, and in-depth interviews with a sample of minority program participants.

5. Interviews were conducted with faculty members who nominated students involved in the study and who subsequently tracked them through to the completion of their BA degree or beyond.

History

The Washington Center was founded in 1975 as an independent nonprofit, educational organization to provide students and their colleges with an adjuct academic program in the nation's capital. Its founders recognized that a single organization in Washington could provide placement services and supervision to students in their internships as well as secure faculty to teach seminars, coordinate speakers' programs and conduct special symposia programs. The provision of housing the participants in a centralized residential facility was incorporated. As a result, scores of colleges and universities could gain access to Washington's virtually unlimited resources. Many institutions now use the Center's programs as an important component of their formal undergraduate curricula such as Grinnell College, Spelman College and Furman University.

The Washington Center's internship and symposia programs have grown more than sixfold in ten years. From fewer than 200 students per year in 1976–77 to over 1,400 per year today. The Center attributes this growth to the fact that unique educational programs appeal to students and faculty from a diverse range of majors, interests, backgrounds and regions. Additionally, as the Center's programs have grown, the number of institutions that endorse or require off-campus internships for credit have also grown. Students and their universities increasingly regard these kinds of experiences as a necessary part of the educational process and have come to depend on the Center to provide them with regular access to those experiences.

Over 10,000 students since 1975 have taken important first steps towards achieving their goals by participating in the Center's programs. During this time period, minority participation also grew dramatically. Between 1975 and 1980 fewer than fifty minority students participated in the program. From 1980 to 1986, over 500 minority students have participated in the program. With the Center's successful programming and forward thinking, it has gained national recognition for its success in giving undergraduates a head start on their professional, academic and civil careers.

Concept

The concept of utilizing community resources for the enhancement of minority student retention developed from recommendations given by college presidents at a conference the Center sponsored in 1980. With funding from a Ford Foundation grant, the Center invited twenty college presidents to attend a roundtable discussion of the Center's programming concerning minority students. Evolving from this discussion were a series of support systems which were developed by the Center.

As the Center implemented its support systems it

gained stature in the D.C. community and its name became more easily recognized for the outstanding work that it provides students and the community. It has also gained financially through the work of the development office. The development office's ability to raise funds from community members has resulted in direct contributions from boards on contacts and access to other funding sources. The Center raised over $1,500,000 in scholarship funds for its students and programs. In a reciprocal fashion, community members gained contacts, networking capabilities, knowledge through programming and marketing possibilities for their community programs. Everyone involved gained from the programs the Center implemented from community resources.

Support Systems & Programs

Mentoring Systems

Some schools have tried or are presently implementing mentoring programs for their students. Almost all of these mentoring programs have been based on two levels. The first, involves students with faculty and administrators as mentors. The second major mentor program involves faculty and or administrators with each other.

The student/faculty mentor programs have failed because black students are already isolated and do not readily take into their confidence administrators or faculty members who intimidate the students. Some faculty and administrators have looked upon such programs as another additional task that may have good intentions but increases their workload. Thus, the intimidation factor for the student is increased because the faculty/administrator cannot provide as much time as the student would like on a wide breadth of issues students are facing.

In order to rectify this problem I would advocate the expansion of mentoring programs to include community professionals. This will increase resource persons to assist students in identifying more specific career goals, future and alternative educational directions, their development in off-campus community projects and their direction in advanced studies.

We usually think of career counselors as providers of information to students, however, we should take advantage of utilizing our community and local professionals to enhance the learning experience of our youth and augment our resource base.

Community assistance for mentors can be derived from:

- Presidents of Businesses
- Store Managers
- Attorneys/Physicians
- School Administrators
- Athletic Teams
- Corporate Directors
- Journalists and Broadcasters
- Politicians
- Artists/Craftpersons
- Retired Professionals

Aspects of mentoring and its purpose:

- To help students become more motivated.
- To allow the students an opportunity to relate to

someone who went through what they are going through.

- To develop more positive attitudes about themselves and their growth experience.
- To help them in relating to and observing the professional who has "made it".
- To help them to more effectively develop their communications systems.
- In turn, the mentor lends his or her sense of expertise to the student.
- The mentor can provide ideas, issues, alternatives, advice and guidance to students.
- The relationship will result in positive benefit to students so that they, in turn, can "mentor" and motivate someone within the next generation.

It is important to examine the structure, format and design of the establishment of a mentoring relationship. Ten questions to consider when establishing mentoring programs are as follows:

1. Should the mentoring program be voluntary for students?
2. How large should the mentoring program be?
3. Should matches be made by majors, career interests or some other area?
4. Should matches be made by race, class or sex?
5. How will the mentoring program be evaluated?
6. How do you determine a mentor's "suitability" to be a mentor?
7. How many students to match with a mentor?
8. How will the matches be made?
9. What format or kind of orientation will take place for the students and mentors?

10. How many students do you begin with i.e., the freshman class, selected majors, rsidential halls, or all minority students?

The role of th mentor and the mentoree is most significant to highlight. Both individuals should share a common purpose and communicate in a mood of openness. As a consequence, the elements of trust and freedom and sense of community are manifested. There is an investment of commitment and often, one must take risks in getting to know a previously unknown individual and commit to the responsibility of the overall mentoring experience. It is important to help students establish self confidence, self awareness and provide them more lucid directions about future academic and career pursuits which will emerge from the mentoring experience.

Once the ten conceptual questions for establishing a mentor's program have been addressed, a two to four page concept paper should follow. This paper should serve as the basis for all other developmental aspects of the program including the following sequential steps:

1. Contact the alumni office for a listing with addresses and phone numbers of all minority alumni within your community.
2. Establish an Advisory Board for the program to meet at least twice before the program begins. The Advisory Board should consist of alumni, community leaders, church leaders and representatives of specific businesses represented in your program. Prior to the first invitational letter for the estabishment of the Advisory Board, it is essential to create a job description listing the roles, terms of office, expectations and time commitments. The

concept paper and job descriptions should be forwarded to the perspective board members.

3. Hold the first meeting and get advice on additional people to contact as mentors. Ask the board members to contact their friends to increase the mentor pool. Write thank you letters to your board members.

4. Mail a job description for becoming a mentor to prospective mentors. List the expectations, roles, and theory behind the mentor programs. Enclose a form to be returned by a certain date including biographical information on each potential mentor. Mail to local alumni and other contacts a list of the board.

5. Send a thank you letter to those who responded positively stating when they should expect to hear from you with a mentor/student and when is the opening reception to meet their students.

6. Have a form developed where students list their career aspirations and any other information you deem important in making a match.

7. Mail these student forms to the prospective mentor with an invitation to the opening reception.

8. Give the students information about the program and a copy of the form you received from the mentor concerning their interest in the program.

9. Hold an opening reception and introduce the student to the mentor. Encourage the mentor to take the lead in the conversation because some students may be intimidated. Also have a form available at the recepiton detailing when the next meeting between the mentor and student will be held. Have the student and mentor sign the form. Retrieve and mail to each as reminders.

10. Forward an evaluation form of the program towards the end of the program to the mentor. You may also want to establish a mailing system or newsletter informing mentors of events on campus that they may wish to meet their student.
11. Evaluate the program and implement systems to improve the program. Thank the students and mentors for their participation.

Positive developments occur as access is provided for students to community leaders/mentors. Mentors and role models for the students are developed as well as opportunities to apply classroom theories to real world experience through discussion. Students should have more access to school and summer employment opportunities and financial and/or scholarship opportunities from various leaders/mentors.

What better way for a student to relate to making it through school than by working with an alumnus? That alumnus can talk to the students about academic and social problems they may have faced and how they dealt with those problems. Alumni also make excellent role models and mentors. Use of local alumni can provide success stories for the students.

In conclusion, the establishment of the community based minority mentoring programs does not have to be a huge expense item in anyone's budget. To establish the program one could have graduate students develop the systems as part of their assistantship. Once the program is established, it would be an excellent opportunity for one or two senior students to coordinate an honors program under the direction of a specific office. It is also vital that the program receive the endorsement of the President and other offices on campus such as development, alumni relations, career planning and others. The president's endorsement would assist in obtaining cooperation among different departments.

Community Based Symposia

So often college students know very little about their communities. By establishing a mentoring system this lack of knowledge will dissipate. Community symposia could enhance the mentor program and involve more students in the concept of community involvement. A key component of the program would be to invite the community to come to campus to hear their community leaders speak. This would increase interaction between the community and the students. The Washington Center is now entering its eighth year of symposia programming on timely national and international issues. Students, faculty and community professionals investigate issues such as foreign affairs, politics, legal careers and leadership roles for women and men through symposia programs. Any subject could be analyzed through this medium.

All such programs give participants direct access to decision makers and political figures who influence public policy through lectures, briefings, site visits, panels and debates. Alumni also have a tremendous opportunity to enhance their prestige in the community by participating in the programs.

Speakers Programs

Community-based speaker programs have provided students tremendous exposure to leaders from a variety of forums. In many instances, students are familiar with national leaders, but community leaders, corporate CEO's, bankers, writers, etc. are rarely approached by colleges and universities to speak to their students.

The Washington Center uses community leaders as well as nationally recognized spokesmen in a variety of ways. Speaker series events are held on a daily, weekly, monthly and semester basis. Subject matter would reflect numerous levels of interests which could be interrelated with your students' interests, whether national, local or college major related. The speaker program should be scheduled on a consistent basis. A minority affairs office, or a multicultural office could coordinate the speaker's program.

The Washington Center produces a Monday Night Lecture Series, Capitol Hill Breakfast Series, Brown Bag Lunch Series and a Community Leaders Forum Series each month. As students learn more about subjects from community leaders, they also develop more confidence in venturing from the campus for more positive reinforcement in activities such as internships, co-ops and employment. As the students venture out, they are also interested in expanding their networking capabilities. Within the college community, students are always meeting new people and developing a campus network. Professional networking opportunities really do not exist in a less informal atmosphere. Job interviews can be arranged for students by the career placement office. Sutdents need to meet with professionals in their fields in more informal settings. Speaker series provide this networking capability.

Adopt-a-Church

Black and Hispanic students sometimes have particularly close ties to the church. Instituting a formal program involving churches creates more access to information and leaders for numerous students. If one involves the churches in its speaker and symposia programs, the college or university's images within the community could be enhanced.

Once again, the minority affairs or college chaplain's office could coordinate this project. Instead of just providing a listing of churches for students, one could take a proactive approach. Contact can be made with those churches that have representative populations equal to those of your minority students. If there are numerous churches perhaps an advisory board could be established.

Advisory Boards

The essence of the advisory boards should be to offer and share useful information for a specific program or mission. The use of such a board can help sustain your commitment and purpose to furthering the educational goals of students within colleges and universities.

Imagine a group of individuals with diverse heritages and backgrounds who can generously share their time by offering their ideas, concerns and strategies for helping black students.

It is important to invite people to serve on the advisory board who:

1. Have a commitment to your students' welfare;
2. Have the time to serve on your board;
3. Can offer stimulating ideas, strategies and know-how to your overall efforts;
4. Offer your program a divese set of role models;

5. Can enhance the visibility or bring significant notoriety to your institution; and
6. Work well together.

Summary

This paper provides a model for mentoring, advisory boards, speaker series, adopt a church, and host family programs. Each model provides a way for colleges and universities to expand or create programs that involve a long neglected area—the community. Many students attend colleges and universities and never know anything about the existing community beyond the limits of the college campus.

A college can be an alien environment, particularly on a predominantly white campus, for minority students. Campuses around the country are seeking ways to increase their retention rates. The models provided in this chapter may be used to increase the retention of black students.

References

Anderson, E., and Hrabowski, F. Graduate school success of black students from white colleges and black colleges." *Journal of Higher Education*, 1977, 48, 294–303.

Branson, H. R. Black Colleges of the North. In C. V. Willie and R. R. Edmonds (eds.), *Black Colleges in America*. New York: Teachers College Press, 1987.

Fleming, Jacqueline, *Blacks in College*, San Francisco, California, Jossey Bass Incorporation, 1984.

Gibbs, J. T. Patterns of adaptation among black students at a predominantly white university. *American Journal of Orthopsychiatry*, 1974, 44, 728–740.

Jones, A. *Uncle Tom's Cabin*. New York: Praeger, 1973.

Jones, M. The responsibility of the black college to the black community: Then and now. *Daedalus*, 1971, 100, 732–744.

Peterson, M. W. and others. *Black Students on White Campuses: The Impact of Increased Black Enrollments*. Ann Arbor, Michigan: Institute for Survey Research, 1979.

Willie, C. V. and Edmonds, R. R. (Eds.) *Black Colleges in America*. New York: Teachers College Press, 1978.

Bibliography for Mentoring

Alleman, E., Corhran, J., Doverspike, J. and Newman, I. (1984). Enriching mentoring relationships. *Personnel and Guidance Journal*, *62*, 329–332.

Cameron, S. W., & Blackburn, R. M. (1981). Sponsorship and academic career success. *Journal of Higher Education*, *52*.

Collins, Eliza G. C., & Scott, Patricia. (1978, July–August). Everyone who makes it has a mentor. *Harvard Business Review*, *56* (4), 89–101.

Daloz, Laurent A. (1986). *Effective teaching and mentoring: Realizing the transformational power of adult learning*. San Francisco: Josses-Bass.

Gerstein, Martin (1985). Mentoring: An age old practice in a knowledge-based society. *Journal of Counseling and Development*, *64*, .156–157.

Glidewell, J. C., Tucker, S., Tody, M., & Cos, S. (1983) Professional support systems: The teaching profession. *In new directions in helping, Volume 3: Applied perspectives on help-seeking and receiving*, pp. 189–212. Edited by A. Nadler, J. D. Fisher, and B. M. DePaulo. San Francisco: Academic Press.

Hennecke, M. (1983). Mentors and protegees: How to build relationships that work. *Training*, 20 (7), 36–41.

Johnson, M. (1980) Mentors—The key to development and growth. *Training and Development Journal*, 34 (7), 55–57.

Kram, Kathy E. (1985). *Mentorizing at work: Developmental relationships in organizational life*. Glenview, IL: Scott, Foresman.

Kram, Kathy E. (1983, December). Phases of the mentor relationship. *Administrative Science Quarterly*, 26.

Lea, D. & Leibowitz, A. (1983). A mentor: Would you know one if you saw one? *Supervisory Management*, 28 (4), 33–35.

Levinson, Daniel J. (1978). *The Seasons Of A Man's Life*. New York: Alfred A. Knopf.

Little, Judith Warren (1985), November). Teachers as teacher advisors: The delicacy of collegial leadership. *Educational Leadership*, 34–36.

Moore, K. M. (1982). *The Corporate Connection: Why executive women need mentors to help them reach the top*. Englewood Cliffs: Prentice Hall, Inc.

Moore, K. M. & Salimbene, A. M. (1981, Fall). The dynamics of the mentor-protege relationships in developing women as academic leaders. *Journal of Educational Equity and Leadership*, 2 (1), 51–64.

Phillips-Jones, L. (1983, February). Establishing a formalized mentoring program. *Training and Development Journal*, 38–43.

Roche, Gerald (1979, January–February). Much ado about mentors. *Harvard Business Review*, 20, 14–28.

Sheehy, Gail (1976). The mentor connection: The secret link in the successful woman's life. *New York Magazine*, April 5, 33–40.

Schmidt, J. A. & Wolfe, J. S. (1980). The mentor partnership: Discovery and professionalism. *NASPA Journal*, 17 (3), 45–51.

Shapiro, E. C., Haseltime, F. P., & Rose, M. P. (1978, Spring). Moving up: Role models, mentors and the patron system. *Sloan Management Review*, 19 (8), 51–58.

The Woodlands Group (1980). Management development roles: Coach, sponsor, mentor. *Personnel Journal*, 918–921.

Zey, Michael G. (1984). *The mentor connection*. Homewood, IL: Dow Jones-Irwin.

Chapter VI

Financial Aid and Tuition: Factors Contributing to the Decline of Black Student Enrollment in Higher Education

Dr. Glenda F. Carter
Coordinator of Research
Grambling State University

Financial aid has come to play a vitally important role in students' ability to gain access to and persist in higher education programs. It has been estimated that one-half of all students enrolled in post-secondary institutions receive some type of financial aid. The percentage is even higher for black students enrolled in black institutions, where over 80 percent at private black colleges (NACUBO, 5/1987), and over 90 percent at public black colleges have, at some point, received federal financial assistance (American Council on Education, 1982).

Since financial aid to students has become a larger share of the revenue generated by colleges and universities, it is more important that questions be asked and answered about the relationship between availability of financial aid and student access to and persistence in college. This paper begins such an inquiry by asking and addressing the question: "What impact have financial aid trends and tuition increase had on the declining enrollment of black students in higher education?"

Purpose and Scope

Despite some gloomy early forecasts, total college enrollment continues to increase. A Department of Education statement reported that 12.39 million students were enrolled in the fall of 1986, compared to 12.25 million in 1985 (NACUBO, 3/87).

The growth is due, in large part, to the increase in students over age 25 who are attending college in larger numbers. This increase in older students attending college has been able to offset the effect of the decline in the population of more traditional aged (18–24) students. Individuals over age 25 now make up 40 percent of the total college enrollment. This trend toward more older students is expected to continue into the early 1990s (United Way, 1987).

Despite the overall increase in enrollments in higher education, black student enrollment in the nation's colleges and universities has declined. Even more baffling is the fact that the number of high school graduates and the graduation rate for blacks is at an all-time-high (Arbeiter, 1987). It is reported that between 1976 and 1985, the high school graduation rate of black students rose from 67 to 75 percent. During the same period, however, black students' college-going rate dropped from 34 to 26 percent (Jaschik, 1987). These facts demand that attention be given to the question of why there is a decline in the number and percentage of black high school graduates who go on to college, when their number and percentage of the population continue to increase. This paper addresses that question from a strictly financial perspective. More specifically, it explores the relationship of declining black student enrollment to decreased availability of financial aid, tuition hikes, and the shift in emphasis from grants as the primary source of aid to loans as the primary source. The paper traces the pattern of black student college attendance along with the patterns and trends related to financial aid. The aim of the paper is to show that major shifts in financial aid trends and tuition hikes coincide with decreased black student enrollment in college. Although no empirical cause-and-effect relationship can be established between the variables, the implications are strong. All indicators point to increased obstacles for black, low-income students seeking access to college.

In addition, this paper highlights ways of minimizing the impact of these obstacles by developing alternative sources of financial aid and innovative tuition payment plans. It is postulated that developing alternative financing strategies is the key to offsetting the deleterious effects of decreased availability of financial aid, tuition increases and changes in the emphasis of aid programs.

Trends in Black Student College Enrollment

Since the peak enrollments during the latter half of the 1970s, black student enrollment has steadily decreased in virtually all attendance categories including, full-time attendance and four-year college attendance (Arbeiter,

1987). United Way (1987) reports that in 1976, 34.6 percent of Hispanic and black high school seniors (combined) attended college. By 1983, that figure had dropped to 29.2 percent. During the same period, the proportion of black high school graduates (alone) enrolling in college dropped from 32 percent to 27 percent (U.S. Bureau of the Census, 1984; Jensen, 1986). Another source reports that in 1980, 1,101,000 black students were enrolled in college; but by 1982, that number had dropped to, 1,094,000 (Arbeiter, 1987).

Table 1 shows the status of black enrollment in relation to other ethnic groups. It reveals that between 1976 and 1984 undergraduate enrollments increased for all racial groups, except blacks. Asians showed the greatest gains with an 86.8 percent increase during this eight-year period. Hispanic enrollment increased by 23.4 percent, while whites showed a 5.7 percent gain and Native Americans, a 4.5 percent gain. In sharp contrast, however, the table shows that between 1976 and 1984, a 3.9 percent decline occurred in the number of black undergraduate students. As previously noted, these statistics are cause for concern given that the pool of black high school graduates has been on the rise.

The degree to which the population of black 18-to 24-year old high school graduates has increased in comparison to other groups is shown in Table 2. The table reflects a 5.1 percent decline (from 26.3% to 21.2%) in the ratio of black undergraduates to the population of black 18-to 24-year-olds. Table 2 further shows that the ratio for whites remained relatively stable with less than one percent variation during the eight-year period from 1976 to 1984. The ratio for Hispanics also declined but to a lesser degree than for blacks.

The statistics verify the stability of the black college-

TABLE 1

Undergraduate Enrollment in Higher Education by Race and Ethnicity: 1976, 1980, and 1984

	Students	White	Total Minority	Asian*	Black	Hispanic	Native** American
Fall 1976	8,432,240	6,899,743	1,402,487	152,533	865,147	323,540	61,267
Fall 1980	8,262,820	7,465,722	1,606,192	214,989	932,055	390,440	68,708
Fall 1984	8,063,178	7,293,747	1,579,267	284,897	830,986	399,333	64,051
			Percent Change				
1976–1980	9.9	8.2	14.5	40.9	7.7	20.7	12.1
1980–1984	−2.2	−2.3	−1.7	32.5	−10.8	2.3	−6.8
1976–1984	7.5	5.7	12.6	86.8	−3.9	23.4	4.5

*Includes Pacific Islanders

**Includes Alaskan Natives and American Indians

Source: W. R. Allen, 1987, p. 36.

TABLE 2

Population 18- to -24-Year Olds—High School Graduates and College Enrollment: 1976, 1980, and 1984

	Total	White	Black	Hispanic
		(Numbers in Thousands)		
1976 population 18-to-24 years	26,936	23,157	3,293	1,437
High School graduates	21,006	18,484	2,131	782
Percent	78.0	79.8	64.7	54.4
Enrolled in college	8,432	6,900	865	324
Percent	31.3	29.8	26.3	22.5
1980 population 18-to-24- years	29,118	24,717	3,711	1,954
High school graduates	22,449	19,469	2,454	1,099
Percent	77.1	78.8	66.1	56.2
Enrolled in college	9,263	7,466	932	390
Percent	31.8	30.2	25.1	20.0
1984 population 18-to-24- years	28,768	23,939	3,919	2,019
High school graduates	22,465	19,028	2,808	1,122
Percent	78.1	79.5	71.7	55.6
Enrolled in college	9,063	7,294	831	399
Percent	31.5	30.5	21.2	19.8

Source: W. R. Allen, 1987, p. 36.

age cohort as indicated by the steady increases in the number of black high school graduates. If there had been a reduction in the black college-age cohort during the latter part of the 1970s and early 1980s, it would have been reflected first in decreased numbers of black high school graduates, which is not the case as shown in Table 2.

The steady, alarming declines in black student college enrollment can, perhaps, be attributed to factors other than rising costs of education and diminishing availability of financial aid. However, few of the variables put forth as possible correlates of deteriorating black student participation in higher education have been supported by research. For instance, growing death rates of black teenagers due to violence has been suggested as a possible reason. The evidence, however, points to a sharp reduction in the black teenage death rate between 1973 and 1983 (Arbeiter, 1987). Neither is the theory that more black youth are in prison fully substantiated. The data show that the arrest rate for both whites and blacks has declined considerably from its high point in 1978, even though the arrest rate for black youth is still approximately twice that for whites. The high birth rate, especially the out-of-wedlock births, is another factor cited as contributing to the drop in black student college enrollment. "This factor is considered critical since women represent a high proportion of blacks going on to college" (Arbeiter, 1987, p. 16). Again, however, the theory is not supported by the data. While white teenagers do have a lower birth rate than black teenagers, the gap is narrowing at a steady rate. This is due largely to the fact that the birth rate for unmarried white teenagers is increasing at a rapid rate while the rate for black teenagers is experiencing a slight decline. Arbeiter (1987) reports that the ratio of black to white teenage birth rate dropped to 6.5 in 1978, 4.7 in 1983, and 4.6 in 1984. These figures would suggest that the teenage birth rate is probably not a significant contributing factor to declining black student college enrollment.

More probable contributing factors are tuition increases and dwindling financial aid resources. Data from student exit interviews at one institution confirm the notion that lack of adequate financial support plays a major role in the premature departure of college students, especially minority students (Arnold, et al, 1986). Other studies have also shown that enrollment rates are positively associated with money spent on student aid (Leslie & Brinkman, 1987); and the escalating college costs along with the current trends in financial aid are likely to contribute further to decreased access and retention.

Trends in Student Financial Aid

Financial aid to college students rose to a record 20.5 billion dollars in 1986–87 (Evangelauf, 1987a). From 1980–81 to 1986–87, student financial aid increased 20.7 percent. These are findings of a recent College Board study. Unfortunately, the Board report also showed that when adjusted for inflation, student aid actually declined 6.1 percent (Lewis & Merisotis, 1987).

In addition, it was found that from 1985–86 to 1986–87, total student aid increased 2.7 percent. Federal aid rose slightly—up by 0.5 percent, while state aid jumped 15 percent and institutional aid climbed by 7.3 percent (Evangelauf, 1987c).

These figures would suggest that more aid is available to students; but this is not necessarily the case. Most student-assistance dollars still come from the federal government, but the contribution to the total pool of available aid has been decreasing. In 1980–81, 83 percent of all aid was supplied by the federal government. By 1985–86, federal monies were 74 percent of the total. Also, during this same period, the American Council on Education (1985) reports that:

1. The value of grant aid plummeted from 40.4 percent to 25.7 percent of total student expenses. Pell Grant

funds awarded to students attending independent institutions fell 34.5 percent in constant dollars.

2. Reliance on loans increased from 12 percent to almost 20 percent of total student expenses over four years. The average annual loan is now $2,256.

3. Students and their parents are paying an increasing share of the costs of attending independent institutions, up from 53 percent in 1979–80 to 65.5 percent in 1983–84 (p. 3).

4. Institutional aid, awarded to three out of every five federal aid recipients, has more than doubled to almost $2 billion annually in the two years between 1981–82 and 1983–84 (p. 3).

Reaganomics and Student Aid

The changes in financial aid outlined above can be attributed primarily to Reagan administration policy changes. Various legislative actions led to the phasing out of programs, changes in eligibility requirements and a dramatic shift in emphasis in the composition of federal aid.

More specifically, stipulations of the **Omnibus Budget Reconciliation Act of 1981** have made Guaranteed Student Loans (GSLs) more difficult to secure and to administer. The Act required that students applying for GSLs show need if their family income exceeded $30,000. It also stipulated that students receiving the loans pay a five percent origination fee on each new loan, i.e., a student had to sacrifice five percent of the loan amount in order to get the remainder of the loan. The interest rates also increased (Doyle & Hartle, 1984). Other legislative changes reduced the maximum amount that students could borrow from $2,625 a year to $2,000. The five percent origination fee will be dropped, but borrowers will be charged a nine percent "guarantee fee" to cover default and other administrative costs.

Other eligibility requirements further complicate matters. Independent student status will no longer be determined by campus administrators, and students under age 30 will have to prove independent status by submitting parental tax forms.

Moreover, the administration's plan reduced the special allowance payment to lenders which had compensated them for the difference between the interest rate they receive on a loan payment and the market rate for borrowing. In addition to receiving less money for a student loan, banks will be forced to take greater risks in offering GSLs because the federal government will no longer pay 100 percent insurance on defaulted loans. Ten percent will have to be absorbed by the lender and loan guarantee agencies (Student Aid News, 1987). The new effect of such action will probably result in limited involvement or perhaps discontinuation of involvement in the GSL program by lending institutions.

The Pell Grant was also affected by the 1981 Reconciliation Act. Provisions allowed Congress to impose a ceiling on grant amounts and insure that the ceiling was not exceeded. Previously, Pell Grant provisions mandated that the government appropriate whatever funds were necessary to serve all eligible students. The ceiling means that money simply will not be available to thousands of students qualified for this aid.

The cuts and modifications to the Pell Grant program demand that families of dependent children pay more of the costs of a college education. Students from a family of four and an annual income below $10,600 are the only ones not expected to make a family contribution. In general, the grants are limited to students from families with annual earnings of $20,000 or less (Student Aid News, 1987).

The unwillingness of the federal government to continue its institutional administrative allowance for Pell Grants poses another threat. Colleges involved in the Pell Grant program will no longer get the $5 per recipient to help defray administrative costs (Student Aid News, 1987). Consequently, Pell Grants will not only become harder for students to obtain, they will also cost colleges more to administer (Carter, 1988).

Another major provision of the 1981 Act altered and eventually phased out the Social Security program. Enacted in 1962, the Social Security program had provided educational funds to students who were dependents or survivors of Social Security beneficiaries (Hartle, 1985; Lee, 1986). The Reconciliation Act of 1981 mandated that no new students were eligible for the funds after June 1982. In 1985, the program was phased out altogether (Lee, 1986).

In 1986, more Reagan Administration policy changes went into effect. **The 1986 Budget Reconciliation Bill** called for a reduction in the federal deficit by $18 billion over a three-year period. As a result, the GSL program will ultimately lose over $806 million in funding over the three-year period (American Council on Education, 1986). Overall, the reconciliation bill seeks a 26 percent reduction ($2.3 billion) in higher education programs from the 1986 appropriation.

The potential far-reaching effects of the reconciliation bill on higher education do not seem as severe, however, when compared to the even more deleterious effects arising from stipulations of the Gramm-Rudman-Hollings deficit-reduction law. Gramm-Rudman-Hollings (GRH) is designed to balance the federal budget by fiscal year 1991, by establishing a set of deficit targets done on an incremental basis. The legislative action provides an avenue whereby automatic spending cuts are imposed if the spending and tax laws on the books do not result in the prescribed deficit target for a given year (Hauptman, 1986).

GRH cuts are based on growth of the economy. Therefore, an economic slump could result in large increases in the size of spending cuts required. GRH is more of a threat to student assistance programs than Reagan policy because it would impose cuts on all aspects of federal support for higher education, whereas Reagan policy proposes to increase federal dollars to some higher education programs (Carter, 1988). (It should be noted that recent amendments to GRH cancelled a low fiscal 1989 deficit ceiling that would have required deep cuts. However, future policy is still constrained by the law.)

Shift in Emphasis from Grants to Loans

In general, the current evidence points toward an era of increased dependence on loans and decreased availability of grants. Over the last 15 years, student borrowing has risen from slightly over $1.5 billion in FY 1970–71 to over $10 billion in FY 1985–86 (Hansen, 1986a).

During this period, the size of an average annual GSL more than doubled, up from $988 in 1970–71 to $2,277 in 1985–86. Moreover. The number of borrowers increased nearly four-fold, jumping from $1,017,000 in 1970–71 to 3,640,000 in 1985–86. Today, one-third to one-half of all students at public four-year colleges graduate with an average of $6,685 in loans; those at private four-year colleges with roughly $8,950 in loans. Even students at two-year colleges accumulate an average debt of $3,000 to $4,500 in loans. The bulk of borrowing has occurred in the GSL, Perkins Loan (NDSL), and Parent Loans to Undergraduate Students (PLUS) programs. In the coming years, the new Income Contingent Loan (ICL) is expected to become a primary source of federal aid. The ICL is suppose to make loan repayment more flexible by tying repayment to income after graduation.

These trends show how needy students are being forced into large loan debt due to the evolution of loan programs away from "minimally subsidized loans of convenience for middle-income students to highly subsidized loans for needy students" (Evangelauf, 1987b, p. 18). The balanced array of grants, loans, and work opportunities that was once the focal point of student aid programs has given way to loans, and disadvantaged students are no longer the focus of federal concern (Hansen, 1986b). Resistance against the shift in emphasis is no doubt reflected in the decreased participation of minority students in higher education.

Trends in College Tuition Costs

Decreasing availability of financial aid is certainly a factor which contributes to student access to and retention at the nation's colleges and universities; and the impact of this situation is only exacerbated by the steady increases in tuition costs. The burden placed on students by these two major higher education trends leaves many students face-to-face with the question of whether or not a college education is within the realm of possibility. (The enrollment statistics, especially for minority students, would suggest that many perspective students decide to forego a college education.) This decision

101

supports the demand theory which purports that demand for a particular product or service is a function of price, the buyer's income, the price of other products and services, and the buyer's personal preferences. When applied to higher education, this theory suggests that there will be a negative association between college enrollment rates and the prices students are charged (Leslie & Brinkman, 1987; Stafford, et al, 1984). However, it has been pointed out that the negative association between tuition costs and enrollment can be minimized if the higher tuition is matched by expanded financial aid, and if the increased aid is effectively targeted to the most financially-needy students (Seneca & Taussig, 1987). This investigation lends credence to the demand theory while at the same time, it supports the contention that decreased financial aid in conjunction with higher tuition costs compounds the problems faced by students who want a college education.

The increases in college tuition costs have been steady and quite significant over the past ten years. The average annual tuition hikes have generally ranged from five percent to a whopping 20 percent and there appears to be no end in sight. In fact, it is estimated that by the time a child born in 1987 graduates from high school, a college education at a private school will cost over $120,000 compared to the nearly $50,000 cost today. This figure is based on the assumption that the annual inflation rate will not exceed five percent (U.S. News, 1987).

Table 3 shows the pattern in tuition costs over the last 10 years. The trend is clear—higher education is becoming more and more unaffordable. During the last seven years, college costs have risen by over 56 percent at two-year colleges and are up by more than 80 percent at private universities (Evangelauf, 1987c).

TABLE 3

Percent of Tuition and Fee Increases at Four-Year Institutions: 1977–78 to 1987–88

	Public	Private
1977–78	0	6
1978–79	5	7
1979–80	6	9
1980–81	4	10
1981–82	16	13
1982–83	20	13
1983–84	12	11
1984–85	8	9
1985–86	9	8
1986–87	6	8
1987–88	6	8

Source: Evangelauf, 1987a

For the seventh consecutive year, college tuition has risen at a greater pace than the rate of inflation. In 1987–88, these increases have ranged from five percent at community colleges to eight percent at independent institutions (NACUBO, 10/87). Quoting from a recent College Board report, and depicted in Table 4, Evangelauf (1987a) reports that:

1. At four-year public colleges and universities, tuition and required fees will average $1,359, up six percent from 1986–87.
2. At four-year private institutions, tuition and fees will rise eight percent, to $7,110.
3. at two-year public colleges, tuition and fees will average $687, up five percent.
4. At two year private colleges, tuition and fees will increase six percent to an average of $4,058 (p.1).

TABLE 4
Average Tuition and Fees and Other Costs at Public and Private Institutions
1987–88

	Public Colleges		Private Schools	
	Resident	Commuter	Resident	Commuter
4-year colleges				
Tuition and fees	$1,359	$1,359	$ 7,110	$ 7,110
Books and supplies	386	386	392	392
Room and board*	2,745	1,228	3,383	1,249
Transportation	380	664	376	632
Other	919	917	721	790
Total	$5,789	$4,554	$11,982	$10,173
2-year colleges				
Tuition and fees	$687	$687	$4 058	$4 058
Books and supplies	673	373	359	359
Room and board*	—	1,244	2,887	1,037
Transportation	—	742	322	609
Other	—	843	679	674
Total	—	$3,889	$ 8,305	$ 6,737

*Room not included for commuter students

—Insufficient data

Source: Evangelauf, 1987a.

Overall, colleges and universities will increase their costs by five to eight percent during the 1987–88 year. The inflation rate in comparison, is expected to be around four percent (Evangelauf, 1987a; NACUBO, 10/87). Table 4 shows how these percentages add up in dollar amounts for 1987–88 college students. It reveals the dollar impact of the five to eight percent tuition raises at four and two-year private and public institutions. It also shows the costs for books and supplies, room and board, transportation and miscellaneous items. According to this study, room and board charges for the fall of 1987–88 rose more slowly than tuition. At four-year public colleges, these charges are averaging $2,745, an increase of four percent. Room and board at four-year private institutions is averaging $3,383, for an increase of six percent (Evangelauf, 1987a; NACUBO, 10/87). At two-year private colleges 1986–87 fall students were met with roughly five percent ($2,887) increases in room and board costs (Evangelauf, 1987a; NACUBO, 10/87).

The projected tuition raises for 1988–89 are illustrated in Table 5. It shows that many institution (33%) expect smaller increases (3 percent or less) in the coming year. Roughly 29 percent of all higher education institutions project tuition raises between six and seven percent, and 21 percent project increases between four and five percent.

Table 5 also reveals the most commonly cited reasons for the tuition hikes. "State-mandated increases in student share of costs" tops the list of cited reasons followed by "catch-up increases in faculty salaries." Other more commonly cited reasons for tuition hikes include "new or expanded academic programs," "fixed costs allocated across fewer students," and "reduced state

TABLE 5
Tuition Forecasts for 1988–89 and Reasons Cited for Increases
(Responses are given in Percentages)

	Type of Institution						
	2-Year	Bacca-laureate	Compre-hensive	Doc-torate	Public	Private	All
10 percent or higher	5	6	6	5	6	5	5
8–9 percent	12	10	14	11	6	19	12
6–7 percent	15	49	34	41	16	44	29
4–5 percent	22	19	20	27	21	21	21
3 percent or less	47	15	26	16	52	11	33%
Reasons given for rate increases							
Reduced state and local funding	37	39	39	42	51	23	38
Computer costs	37	38	45	48	30	50	39
State-mandated increases in student share of costs*	60	67	73	69	64	—	64
Construction and renovation costs	29	31	33	33	15	48	31
"Catch-up" increases in faculty salaries	49	71	46	59	33	81	55
Increased student-aid expenditures	28	61	45	42	14	71	41
New or expanded academic programs	40	51	39	36	30	58	43
Fixed costs allocated across fewer students	43	41	25	9	30	46	37

Note: Results are based on a survey of top administrators at a nationally representative sample of 456 colleges and universities. The response rate to the survey, which was conducted in the spring of 1987, was 82 percent.
*Public institutions only.

Source: Evangelauf, 1987a

and local funding'' (Evangelauf, 1987a; NACUBO, 10/87).

Other higher education officials contend that it is a decrease in federal student aid that has driven tuition up (NACUBO, 11/87). Colleges and universities are said to have responded to the drop in federal aid by increasing their assistance and passing that cost on to the students who can afford to pay more.

Whatever the reasons, the trend is obvious. A college education is quickly moving out of the reach of untold numbers of poor and minority individuals who want to go to college. It is, therefore, imperative that alternative tuition-financing strategies be explored and quickly put into place.

Consequences of the Trends

The decline in black student enrollment which began in the late 1970's can be expected to continue into the next decade, if the financial outlook does not improve. More aid in the form of grants, work-study, and lower tutition is needed to stimulate increased enrollment and retention (Jackson & Weathersby, 1975). Unfortunately, this is not the emerging trend. It is an increased dependence on loans in conjunction with rising college costs that are creating a scenario which limits black access to and enrollment in higher education.

All current indicators point to a positive correlation between availability of financial aid and black enrollment in higher education. At the same time that federal campus-based aid programs have undergone significant reductions, there has been a corresponding decrease in black student enrollment even though overall enrollments have been stable or increasing. Table 1 shows that there was a 10.8 percent drop in black student enrollment between 1980 and 1984. The Omnibus Budget Reconciliation Act of 1981 brought some sweeping changes to federal student aid programs and unfavorably affected two major programs that minority students had come to rely on quite heavily—the Social Security program and the Pell Grants. Grant aid was further decreased during this period due to the reductions in Vietnam-era veteran educational benefits which were limited to use within ten years of military service (Lewis & Merisotis, 1987). During these same years (1980–1984), there were steady increases in the costs of attending college. Table 6 illustrates these patterns in college costs, available aid, and black student enrollment from 1980–81 to 1986–87. As college costs have increased and grants and work-study aid have decreased, and dependence on loans has increased, there has been a corresponding decrease in black student enrollment. Table 6, therefore, supports the reasonableness of the assertion that, indeed, the availability of financial aid and tuition costs impact upon college attendance of black students. While it is conceded that there are many reasons for going or not going to college, this paper argues from an economic standpoint emphasizing the fact that a college education is becoming more expensive and student aid has shifted its emphasis and become less appealing to lower-income individuals, a large portion of whom are black. The overall result is an erosion of the college participation rate of black students.

Alternative Sources of Financial Aid

Inflation, recession, budget deficits, and Reagan policy have all pulled funding away from higher education,

108

TABLE 6
Cost of College Attendance—Total Available Aid and Black Student Enrollment 1980–81 to 1986–87

| | Constant 1982 Dollars (Adjusted for Inflation) | | | | | | | | |
| | Cost of Attendance* | | | Total Available Aid (in millions) | | | Enrollment (in millions) | | |
	Private University	Public University	Public Two-Year	Grants	Loans	Work	Total	Black	% Black
1980–81	5850	2697	2251	10486	7754	736	12,087	1107	9.2
1981–82	6101	2770	2274	9592	8093	640	—	—	
1982–83	6533	2980	2349	8372	7322	604	12,388	1101	8.9
1983–84	6869	3115	2403	8070	8040	648	12,162	—	
1984–85	7163	3210	2562	8115	8696	589	12,233	1076	8.8
1985–86	7538	3326	2669	8482	8682	582	12,300	—	
1986–87	7867	3418	2741	8452	8794	574	12,398+	1066	8.6
% change 1980–81 to 1986–87	+34.5	+26.7	+21.8	−19.4	+13.4	−22.0			

*Cost of attendance includes tuition, fees, room and board.
—Unavailable
+ Estimated

Sources: College Board, 1987. p. 11.
U.S. Department of Education, 1987.
Chronicle of Higher Education, July 23, 1986.

highlighting the need to seek new sources of revenue. Some areas in which colleges have begun to step up their efforts to secure additional funding include: (1) endowment building, (2) procurement of gift funds, and (3) establishment of alliances with lending institutions and other support institutions.

The urgent need associated with the financial problems has given rise to some innovative alternatives in these areas. Some of which are outlined below.

Endowments

In 1983, the Charles S. Mott Foundation in conjunction with the Robert Moten Memorial Institute and the United Negro College Fund began to explore avenues by which historically black colleges and universities (HBCUs) could build endowments. The "Endowment Funding Plan" was set up to provide HBCUs with a substantial grant to be used to leverage a loan that could be as much as three times the grant. Proceeds from both were invested and returns are being used over a long term to pay off the loan and maintain the original investments as an endowment for the institution. This venture was aimed primarily at decreasing dependence on public and private sources for short-term support. Ten HBCUs began experimenting with the plan in 1983 (Spratling, 1983; Carter, 1983). The results have been impressive. The program is still in operation and the original 10 institutions have increased to 37. Despite a slowdown due to economic conditions, the program is expected to expand and offer some financially rewarding opportunities to black colleges.

Gift Funds

The procurement of additional gift funds will be vitally important to colleges and universities in the coming years. Aggressiveness in seeking gift funds from alumni, corporations, churches, local communities, social organizations, as well as national and regional non-profit funding agencies and foundations must become part of the routine financial management practices of higher education institutions.

Many foundations and corporations will assist institutions in this effort by teaching fund-raising strategies and in some instances, will match contributions raised by the institution. Churches and alumni are also valuable assets which must be tapped for support.

Innovative Tuition Payment Plans:
Changing the Psychology of Paying for College

Students and their families are being forced to take on greater responsibility for the cost of attending college. As previously mentioned, the family share of the costs of attending independent institutions was up from 53 percent in 1980 to 63.5 percent in 1984. Current trends suggest a continued move in this direction. Therefore, state and federal governments and colleges and universities are becoming committed to changing the pyschology of paying for college. That is, families are being encouraged to de-emphasize external funding sources when planning for college and instead, plan for the contingency that they will have to produce the greater share of funds needed to finance a college education. This "change-the-psychology" campaign has led to some creative financing strategies. Some of which are discussed below.

111

It appears that tuition payment plans will become quite popular in the coming years. Virtually every state is exploring ways to allow parents to pay their children's college tuition years before the children actually enroll. Governors in at least five states—Indiana, Maine, Michigan, Tennessee, and Wyoming—have signed such plans (Mooney, 1987).

The Michigan Education Trust Tuition Prepayment Plan seems to be the model plan for other states. This plan allows parents or anybody else to pay college tuition costs in advance by purchasing a certificate redeemable for four years' tuition and fees at any of the state's 15 public four-year colleges and universities or any of 29 community colleges (Chronicle, 2/87).

Prices of these certificates will vary depending on the child's age. Lump-sum payments, payroll deductions or installment plan payment options give parents some purchasing leeway. The state will use the payments to set up a trust fund and ultimately use the principal and earning to compensate the institutions when the certificates are redeemed (NACUBO, 2/87).

The Wyoming Tuition-Guarantee Plan makes it possible for parents to lock in the current tuition and room and board costs by setting the money aside now. This program applies to parents of children who are from nine to eighteen years away from college enrollment and it guarantees four years' tuition and room and board at the University of Wyoming or two years at any of the state's community colleges. The eventual savings to be realized are astounding. To illustrate, a child's college education could be purchased today at a lump sum rate of $5,114—$8,806 (nonresident). However, by the year 2003 four years of college at the University of Wyoming is expected to total $24,272 for state residents (U.S. News, 1987). The Wyoming plan assumes that long-term investment of the pool of money received will yield a higher return and provide the necessary funds to cover the escalated costs.

The major concern about and possible drawback of the prepaid tuition plans revolves around the uncertainty over how the Internal Revenue Service will handle the gain in value of the prepayment. Michigan has stipulated that its program will not go into effect if the IRS decides to tax the increase in value of the prepurchased education (U.S. News, 1987; Chronicle, 2/87).

Table 7 shows the number of states with signed prepayment bills, bills pending, or that have prepaid tuition plans under study.

Prepaid Tuition Plans at the Institutional Level

Many colleges and universities are not waiting for state-level tuition prepayment plans to surface; but are developing their own institutional plans. At present, the array of such plans is impressive.

One prepayment plan initiated by the University of Southern California (USC) allows students and their families to avoid inflation by prepaying in full a four-year educational program at tuition rates in effect the previous academic year. Students and their families benefit by avoiding increasing tuition costs. The USC plan allows two options for prepayments—cash payment or loans. The loan program, arranged at local banks, includes: (1) homeowner equity loans, which allow California homeowners fifteen years to repay the tuition prepayment, and (2) unsecured loans, which allow qualified borrowers

TABLE 7

Action by States on Prepaid Tuition

Plans under study*	Bills Pending	Bills signed by governor
Arizona	California	Indiana
Arkansas	Florida	Maine
Colorado	Illinois	Michigan
Connecticut	Missouri	Tennessee
Georgia	New Jersey	Wyoming
Louisiana	New York	
Maryland	Ohio	
Massachusetts	Oregon	
Minnesota	Pennsylvania	
Montana	South Carolina	
Nevada	Texas	
New Hampshire	West Virginia	
New Mexico		
North Carolina		
North Dakota		
Rhode Island		
Utah		
Virginia		
Wisconsin		

*By lesiglative committees or higher-education study committees.
Source: Mooney, 1987 and U.S. News, 1987.

seven years to repay the tuition prepayment (Saurman & Riccucci, 1983).

At Washington University, families can freeze tuition at the freshman level by paying the entire cost upon initial enrollment—often with the help of a fixed-rate loan from the school and as long as ten years to repay. The University of Pennsylvania's "Penn Plan" offers a variety of prepayment and installment-plan options geared to the financial circumstances of all students in the University. Many colleges help families by accepting

114

payment of a year's tuition and room and board charges in monthly installments, usually over a 10-month period (U.S. News, 1987). Other colleges allow families to arrange installment plans through outside financial services (Schwartz, 1986).

Duquesne University permits its alumni and their relatives to pay now for a future education. Interested individuals can purchase today a college education to start in 1999 for $5,593. The university invests the money in zero-coupon U.S. Treasury bonds with maturities matched to the years for which tuition has been purchased. However, if the student chooses not to enroll, Duquesne refunds only the initial investment (Hansen, 1986a).

A similar program at Calvin College sells "Tuition gift certificates" that pre-purchase a college education. The certificates are sold by units, each representing one one-hundredth of a full-year's tuititon at the time of purchase. The university insures that the relative value of the unit will remain intact as tuition increases. Calvin offers no refund of investments, but allows the purchaser to redesignate the certificate to another student if the first designated student does not enroll (Hansen, 1986a).

Linfield College in McMinville, Oregon, will assist families struggling to pay tuition by lending them up to $4,500 over a student's sophomore, junior and senior years at a nine percent interest rate, but the college forgives one third of the principal if the first two thirds, plus interest, is repaid as contracted (U.S. News, 1987).

Schwartz (1986) discusses some tuition payment plans that resemble department stores clearance sales. For instance, Goucher College (Baltimore) offered to each of two students a year's tuition, and room and board at the 100-year-old price of $100 to celebrate its centennial.

Fairleigh Dickinson (New Jersey) has a "two-fer" plan which allows a second sibling to attend at half price if the first attends at full price. Similarly, Lake Erie College in Painesville, Ohio, has allowed twins to attend the school for the price of one. Bard College in Annandale-on-Hudson, New York, is set on attracting the 'cream of the crop' students (top 10 percent of their class) by allowing them to matriculate at the institution for whatever they would pay to go to one of their state colleges or universities.

The University of Minnesota has developed another approach to helping its students pay the costs of tuition. This university reserves all non-academic positions that are 29 hours per week or less for student workers who are paid salaries comparable to those of non-students for the same kind of work. Approximately one third of the institutions work force is made up of students. This program provides job opportunities for over 17,000 students annually (Hansen, 1986a).

Some of the Ivy League schools help their students by encouraging them to become campus entrepreneurs. These institutions sponsor "student agencies" which allow students to sell a variety of products and services. The scope of the program is limited only by the imagination and business finesse of the students (Hansen, 1986a).

The various tuition payment plans discussed here may not be the answer for all families, but they do provide parents with some alternative ways of financing a college education. Encouraging parents to save for college is another approach to paying for the cost of college.

Savings Plans

Perhaps more than ever before, parents are being encouraged to save for their children's college education.

For example, the state of Tennessee sponsors "financial-aid nights" for high school students and their parents. Financial planning experts are brought in for these nights to show parents how saving even small amounts each week can compound over time. The aim of such programs is to encourage parents to start saving early and to "change the psychology of paying for college" (Evangelauf, 1988) from that of outside aid to self help.

Other savings incentive plans would suggest that lawmakers as well as college administrators actively support this campaign. For instance, two key U.S. Senators (Edward Kennedy and Claiborne Pell) have introduced one bill that would help parents save for their children's college education (*Chronicle,* 11/4/87). The *Chronicle of Higher Education* reports that this bill would provide tax breaks for interest earned on U.S. savings bonds that are used to pay for college tuition and fees.

Other lawmakers have introduced bills that provide for the establishment of a national education savings bank, but the Kennedy-Pell Bill is the first that involves U.S. savings bonds. Under this bill it is estimated that parents of a child who is to begin college in the year 2005 would have to save $470 annually in order to pay tuition for four years at a public college. Parents who wish to send their child to a private institution would have to save $2,440 annually (*Chronicle,* 11/4/87).

The Kennedy-Pell Bill stipulates that only parents who earned less than $150,000 a year would be eligible for the tax breaks. In the event that the child decided not to go to college, parents could use the bonds for other expenses, but the interest earned on the money would be taxed (*Chronicle,* 11/4/87).

Additional kinds of savings plans are being explores on a smaller scale. In New Jersey, the College Savings Bank in Princeton has proposed a plan that would allow parents to save for their children's education through certificates of deposit guaranteed to pay interest based on annual increases in college costs (*Chronicle,* 9/9/87). The program is called the College Sure Plan.

This plan allows parents to consult with bank officials in calculating the amount of money they will need to contribute to their savings plan each year. These calculations are based on the type of college they expect their child to attend along with projected annual tuition hikes. The major drawback of the plan is that parents will have to pay taxes on the interest earned by the College Sure savings.

The state of Missouri is also looking at ways to help parents prepare to accept more financial responsibility for their children's education. The state's governor has proposed that parents be allowed to set up special savings accounts for college costs. Contributions to the accounts and the interest earned would be exempt from Missouri income tax (Jaschik, 1987). Families could save up to $2,000 a year without tax liability. In light of the new tax laws, a plan of this nature provides incentive to participate because of the immediate savings on federal taxes. The tax savings, however, occur only if the accounts are in trusts in the student's name, not the parents. The state of Washington has also adopted this IRA-type tax-free savings account (*U.S. News,* 1987).

In Illinois, the governor has approved legislation that would allow state residents to invest in education savings bonds which are created from state municipal bonds. The earnings from these investments would be free of both state and federal income taxes. Also, an interest-rate premium would be earned by those who actually used the funds for college. (*U.S. News,* 1987). An added bonus related to the Illinois savings bonds is the state's assurance that $25,000 of a family's assets will not be considered in calculating a student's eligibility for state scholarships if the money is held in the state's new College Savings Bonds (Evangelauf, 1988).

Other such plans are sure to follow as economic conditions worsen, but until that time, parents will have to employ their own financial savvy in order to pay for their children's education.

Alliances With Lending Institutions to Develop Inventive Loan Programs

In conjunction with the prepayment and savings plans, it is also important that colleges and universities establish any alliances possible to help close the gap between federal and state assistance and the individual student's ability to finance his/her education. One avenue which might be pursued involves negotiating with lending institutions to devise innovative kinds of loans, scholarships and other forms of financing. One example of a creative approach to alternative funding has been tried by Beloit College in Wisconsin—a cancellable loan program. This program establishes a revolving fund in which corporations donate funds to institutions of post-secondary education as a means of retiring their employees' outstanding student loans and attracting graduates to their corporations (Saurman & Riccucci, 1983). The potential benefits to the corporations presumably will elicit their participation.

Yet another alternative approach to providing financial assistance to students would call for a collaborative relationship between the college or university and the municipality in which the university is located. General obligation bonds, backed by a "full faith and credit pledge" could be issued to provide student loans to local residents. Here, the local residents would have the opportunity to approve of the bond sale. However, once a relationship was established between the institution and the municipality, the incentive to students would be to attend hometown institutions as these loans would not be available to out-of-state students (Saurman & Riccuci, 1983).

The state of Illinous reported tremendous success with its first sale of College Savings Bonds. Roughly $93 million was raised for the state with the sale of the zero-coupon bonds. The bonds are exempt from federal and state taxes and provide a cash bonus if they are redeemed for college expenses (Evangelauf, 1988).

These are only a few of the kinds of programs that can be patterned or considered with some variations. Other organizations with which financial packages might be negotiated include businesses, sororities, fraternal groups, community groups and churches. Creativity will be an essential component in developing colleges resources to the point required in order to survive the current attack on higher education by the Reagan administration (Carter, 1985).

Suggestions to Parents

U.S. News (1987) has made several suggestions to parents seeking ways of financing their children's college education. First, it is suggested that parents get one of the guides to college financing. The most popular ones are: *Don't Miss Out: The Ambitious Student's Guide to Financial Aid* by Anna and Robert Leider and *College Check Mate: Innovative Tuition Plans That Make You a*

Winner by Patricia Goeller. The second recommendation is to shop around diligently for tuition help and the best tuition buys. The final recommendation is to become thoroughly familiar with savings options. Fidelity Investments, a Boston-based company has published a 16-page brochure on their "Fidelity College Investment Plan." This brochure gives parents an overview of various investment strategies (Evangelauf, 1988). The College Board has also published a guide on financial aid strategies and instruments for parents entitled *How to Pay for Your Children's College Education* by Gerald Krefetz. Some of these strategies and instruments as outlined in *U.S. News* (1987) include:

1. Buying life insurance. . . . The cash value of a [life insurance] policy grows on a tax-deferred basis while the policy is in force, and you can borrow from the policy at low rates. Furthermore, the cash value of life insurance doesn't count as part of your assets when the time comes to apply for financial aid (p. 87).

2. Using the age of your child to your advantage. If your child is very young, go for growth. . . . no load growth mutual funds have a record of beating the market. . . . Growth funds stress long-term appreciation instead of dividends, so yearly income taxes stay low (p. 87).

3. Checking out savings plans available at work. A 401(k) retirement plan that allows you to borrow from it offers two major advantages. First, the income set aside—up to $7000 each year, tax deferred—won't count among your assets when the time comes to determine your eligibility for financial aid. Secondly, when you borrow from your

account for college, you pay the interest back into your own account (p. 87).

4. Buying U. S. Savings Bonds. As long as bonds are held for at least five years, you'll get a guaranteed interest rate—now six percent. Hold them longer and your rate will float with Treasury security rates but will always be at least the guaranteed rate. Earnings aren't taxed until the bonds are cashed or mature. Zero-coupon bonds—which like savings bonds are bought at a deep discount to face value and pay no actual interest until maturity—generally pay a higher yield. But you'll owe taxes on the imputed interest each year, and you risk a chunk of your principal if you have to cash in before maturity (p. 87).

5. Hiring your kids if you are self-employed. The company gets a tax deduction for their earnings, and a healthy portion of those paychecks goes into savings. This year, a child could earn $2540 before owing taxes; next year this will increase to $3000.

6. Considering a home equity loan or line of credit if you have to borrow when the time comes. As long as the funds are used for education, interest paid on any amount up to the home's market value is fully deductible. Shop carefully, however. Interest rates on most credit lines float upward with market rates. With your house as collateral, you stand to lose in a big way if payments grow too large to meet (p. 87).

Other sources of information on financing options include:

1. *Working Your Way Through College: A New Look*

122

at an Old Old Idea by Pamela Christoffel, Washington D.C.: The College Board, 1985.

2. *The College Cost Book, 1985–86* provided by the College Scholarship Service, New York: The College Board, 1985.

3. *Mortgaged Futures: How to Graduate from School Without Going Broke* by Marquerite Dennis, Washington, D.C.: Hope Press, 1986.

4. *Education at Work: Productivity Through Student Employment,* The National Association of Student Employment Administrators, P. O. Box 1428, Princeton, NJ 08542.

5. *Financial Aids for Higher Education* edited by Oreon Kessler, Dubuque, IA: Wm. C. Brown Company, updated biennially.

In the final analysis, it can be concluded that it will take careful planning, considerable creativity and resourcefulness to manage the cost of a college education. But the good news is, there are ways of doing it.

A Final Word

Fiscal year 1989 promises to bring some increases to student aid programs. In a dramatic turn of events, President Reagan has listed education spending as one of his top priorities. If Congress approves the Administration's plan, a record $8.8-billion will be allocated for student aid (Wilson, 1988). The Administration has supposedly taken a "second look" at the programs which received deep cuts last year after being termed "unnecessary institutional aid." This year, after its reassessment, the Administration views such programs as Supplemental Educational Opportunity Grants (SEOG) and College Work-Study as "important sources of support for financially disadvantaged students" and has asked that money for them be restored (Wilson, 1988).

Other proposals of the fiscal 1989 package include:

1. an increase in the maximum Pell Grant from $2200 to $2300 a year.
2. a provision that students who default on loans or have trouble paying them back be required to have cosigners in order to borrow under the GSL program.
3. a reduction in aid to non-historically black institutions that are in poor financial condition.
4. required outcomes measures to show student success and justify increased campus-based financial aid, i.e., work-study and SEOG funds.
5. an incentive plan to encourage parents to save for college by waiving taxes on the interest accrued from the purchase of savings bonds used to pay for college.

At first glance, it appears that students may get some aid relief *if* the fiscal 1989 package gets the stamp of approval from Congress. However, there also appears to be a number of proposals which will negatively impact on poor and minority students and predominantly (not historically) black colleges and universities. First of all, the GSL cosigner proposal is likely to result in fewer students being able to secure the loans if cosigners cannot be found. Secondly, reduced funding to colleges and universities in poor financial health will disproportionately impact upon predominately black colleges that have a history of financial instability and do not have the status of historically black colleges, as these are not affected by the proposal. Thirdly, even if the 1989 package is approved and student aid is increased by almost $9 billion, when inflationary adjustments are made, we may find as we did in 1987, that student aid actually decreased. And finally, loan programs still remain at the forefront of student assistance packages.

In the final analysis, the best conclusion that can be drawn is that the fate of students assistance programs is uncertain. However, there are some *certainties* with which we must contend now. Specifically, black student college enrollment is declining; the cost of attending college is escalating rapidly; and there has been a major shift in the emphasis of aid programs away from grants to loans as the primary source of student financial assistance. All of these factors point to decreased access to college for black students.

Combating these factors will require creative insight into reducing over-dependence on federal and state support. More pointedly, the financial crisis in higher education will demand that alternative financing strategies be developed to create and maintain equal access and to bridge the gap between available aid and student need.

References

Allen, W. R. Black colleges versus White colleges: The fork in the road for black students. *Change,* May–June 1987, 28–39.

American Council on Education. Campus presidents lobby congress. *Higher Education and National Affairs,* 1986, 35, 7, 1, & 12.

American Council on Education. The Reagan administration's student aid budget cuts: Facts vs. fiction. Washington, D.C., March 1, 1982.

Arbeiter, S. Black enrollments: The case of the missing student. *Change,* 1987, May/June 19, 14–19.

Arnold, L. Mares, K. R. & Calkins, E. V. Exit interviews reveal why students leave a BA–MD degree program prematurely. *College and University,* 1986, 62, 34–47.

Carter, G. F. Financial aid and black students. NAFEO Research Report Number 2, 1988.

Carter, G. F. Linkages, pipelines, and coalitions: Survival strategies for historically black colleges and universities. Paper presented at the Society for College and University Planning Annual Conference, Chicago, Illinois, July, 1985.

Carter, G. F. Mission of historically black colleges and universities: reassessment and redirection. Preliminary examination. Center for the Study of Higher Education, The University of Michigan Ann Arbor, July, 1983.

Chronicle. Kennedy, Pell offer tuition savings bill. *Chronicle of Higher Education,* November 4, 1987, A 37.

Chronicle. Michigan swamped by calls on tuition. *Chronicle of Higher Education,* Feburary 4, 1987, 33, 21, 17.

Chronicle. Racial and ethnic makeup of college and university enrollments, *The Chronicle of Higher Education,* July 23, 1986, 32, 21, 32–34.

Chronicle. Savings plan guaranteed to cover college cost. *The Chronicle of Higher Education,* September 9, 1987, 34, 2, A3.

Doyle, D. P. & Hartle, T. W. Idealogy, pragmatic politics and the education budget. In J. C. Weicher (Ed)., *Maintaining the safety net: income redistribution programs in the Reagan Administration.* Washington, D.C.: American Enterprise Institute, 1984.

Evangelauf, J. Experts extol anew the advantages of saving for college. *The Chronicle of Higher Education,* January 20, 1988, 34, 19, 1 & A32.

Evangelauf, J. Increases in college tuition will exceed inflation rate for seventh straight year. *Chronicle of Higher Education* August 12, 1987a, 33, 1 & 32.

Evangelauf, J. Student financial aid reaches $20.5–billion. *The Chronicle of Higher Education,* December 2, 1987c, 34, 14, A33–A36.

Evangelauf, J. Students' borrowing quintuples in decade, Raising the specter of a "debtor generation." *The Chronicle of Higher Education.* 1987b, 33, 17, 1 & 18.

Hansen, J. S. Alternatives to borrowing. *Change,* May/June, 1986a, 18, 3, 20–26.

Hansen, J. S. Student Loans: Are they overburdening a generation? Washington, D.C. Joint Economic Committee, December, 1986b.

Hartle, T. W. Federal student aid: Where we have been: where we are. In V. A. Hodgkins (Ed)., *New Directions for Institutional Research, Impact and Challenges of a Changing Federal Role,* San Francisco: Jossey-Bass, March, 1985.

Hauptman, A. M. Gramm-Rudman-Hollings: Its potential impact on higher education. *Higher Education and National Affairs,* 1986, 35, 7, 5–8.

Jackson, G. A.& Weathersby, G. B. Individual demand for higher education: A review and analysis of recent empirical studies, *Journal of Higher Education,* 1975, 46, 623–652.

Jaschik, S. College outlook grim for blacks 25 years after barriers fell. *Chronicle of Higher Education,* Sept. 2, 1987, 33, A88.

Jaschik, S. Missouri proposal would encourage parents to save for college fees. *The Chronicle of Higher Education,* 1987, 33, 21, *15* 17.

Jensen, E. L. Assessing recent changes in federal financial aid policy: What will happen to students. *College and University,* 1986, Fall, 62, 24–33.

Lee, J. B. Beyond the pale: how student aid cuts hurt access. *Educational Record,* Spring-Summer, 1986, 20–24.

Leslie, L. L. & Brinkman, P. T. Student price response in higher education. *Journal of Higher Education,* March/April, 1987, 58, 181–204.

Lewis, G. L. & Merisotis, J. P. Trends in student aid: 1980–1987, Washington, D.C.: The College Board, November, 1987.

Moline, A. E. Financial aid and student persistence: An application of causal modeling. *Research in Higher Education,* 1987, 26, 130–147.

Mooney, C. J. Legislatures' financial support for colleges limited by

127

economic conditions in states. *The Chronicle of Higher Education,* June 10, 1987, 33, 39, 23–24.

NACUBO. Enrollment increases one percent. *National Association of College and University Business Officers.* March 1987, p. 14.

NACUBO. Financial pressures on campus continue to build. *National Association of College and University Business Officers,* October, 1987, p. 13.

NACUBO. Increases in tuition outpace inflation for seventh straight year. *National Association of College and University Business Officers,* September 1987, p. 14.

NACUBO. Less aid, not more, drives tuition up. *National Association of College and University Business Officers.* November, 1987, p. 12.

NACUBO. Michigan legislature passes plan for tuition "futures." *National Association of College and University Business Officers,* February, 1987, p. 13.

NACUBO. Students at black colleges hurt by grant cuts. *National Association of College and University Business Officers,* May 1987, p. 16.

Saurman, F. S. & Riccucci, W. M. What are the alternatives to financing students in higher education during a period of retrenchment? *The Journal of Student Financial Aid,* 1983, 13, 2, 35–43.

Schwartz, J. Pay now, learn later. *Newsweek,* April 7, 1986.

Seneca, J. J. and Taussig, M. K. educational quality, access, and tuition policy at state universities. *Journal of Higher Education,* January/February, 1987, 58, 25–37.

Spratling, C. Proud heritage on hold, black colleges: The struggle ahead. Detroit Free Press, May 8, 1983.

Stafford, K. L., Lundstedt, S. B., and Lynn, A. D. Jr. Social and economic factors affecting participation in higher education. *Journal of Higher Education.* September/October, 1984, 55, 590–607.

Student Aid News. Student aid, other higher education programs hit hard in Reagan budeget. *Student Aid News,* 1987, 14, 1, 1–2.

United Way. *What Lies Ahead: Looking Toward the '90's.* United Way of America, Strategic Planning Division, 1987.

U.S. Bureau of the Census. *Statistical Abstract of the United States: 1985.* Washington D.C.: U.S. Government Printing Office, 1984.

U.S. News and World Report. Paying for college, October 26, 1987, 83–87.

Wilson, R. Reagan seeks a record $8.8-billion for aid to students. *The Chronicle of Higher Education,* February 24, 1988, 34, 24, A23–A26.

NAFEO

The National Association of Equal Opportunity in Higher Education (NAFEO), founded in October, 1969, was formed as a voluntary, independent association by historically and predominantly black colleges and universities. It is organized to articulate the need for a higher education system where race, income, and previous education are not determinants of either the quantity or quality of higher education. This is an association of those colleges and universities which are not only committed to this ultimate goal, but are now fully committed in terms of their resources, human and financial, to achieving that goal.

The Association proposes, through collective efforts of its membership, to promote the widest possible sensitivity to the complex factors involved in and the institutional commitment required for creating successful higher education programs for students from groups buffeted by racism and neglected by economic, educational and social institutions of America.

To achieve this goal, NAFEO has determined the following priorities:

1. To provide a unified framework representing historically black colleges and similarly situated institutions in their attempt to continue as viable forces in American society;

2. To build the case for securing increased support from federal agencies, philanthropic foundations and other sources;

3. To increase the active participation of Blacks in the leadership of educational organizations together with memberships on Federal boards and commissions relating to education; and

4. To provide detailed, continuing yearly analyses of constructive information about these colleges and to use that information to help the public develop and maintain a sensitivity to the overall needs of these institutions of higher education.

NAFEO's aim is to increase the flow of students from minority and economically deprived families, mostly Black, into the mainstream of our society.

In carrying out its four major specific objectives, NAFEO serves as a—

1. Voice for Historically Black Colleges

2. Clearinghouse of Information on Black Colleges

3. Coordinator in Black Higher Education

4. Presidential Resource.

The National Association for Equal Opportunity in Higher Education represents the historically and predominantly black colleges and universities of this nation.

132

There are some 117 NAFEO institutions consisting of private 2-year and 4-year institutions, public 2-year and 4-year institutions, as well as graduate and professional schools located in fourteen southern states, six northern states, four mid-west and western states, the Virgin Islands and the District of Columbia. These institutions enroll upwards of 250,000 students and graduate more than 40,000 students annually with undergraduate, graduate and professional degrees. Since 1966, these institutions have awarded a half million undergraduate, graduate and professional degrees. They are the providers of equal educational opportunity with attainment and productivity for thousands of students.

NAFEO'S RESEARCH ADVISORY COMMITTEE

135